W0234524

Resilience to Climate Change

'Drawing on interviews with stakeholders from academic, business, government and not-for-profit organizations in the UK this book provides important new insights on how local resilience to nexus shocks and climate impacts is a complex process of collaboration, communication and adaptation. It brings together key recommendations to inform better decision-making and should be read by academics and policymakers alike. The book should provide an excellent basis for further scholarly work on this extremely important topic.'
—Professor Alison Anderson, *University of Plymouth, UK*

'The challenge of coping with uncertain climate risks and potential disruption across a range of sectors is a topic that is increasingly coming to the fore, with practitioners from the environment, infrastructure, government, and other disciplines looking for solutions. In this work, Dr. Howarth and colleagues skilfully set out the nature of the problem, the barriers to responding effectively, and some innovative thinking about the importance of collaboration and communication to improve resilience.'
—Kristen Guida, *London Climate Change Partnership, UK*

'Understanding the complex ways in which the 'Nexus' of water, food, energy and wider environmental risks can generate shocks to our social and economic systems is a vital task. Candice Howarth and her co-authors have provided a valuable and admirably clear guide to the challenges of 'Nexus shocks' and how decision-makers can face up to them. Anyone concerned with the communication of environmental risks, public understanding of science and risk and planning for resilient responses to environmental shocks to our systems, will benefit from reading this book and applying its lessons.'
—Dr. Ian Christie, *University of Surrey, UK*

'Planning for and responding to climate shocks in our highly interconnected world means we can no longer think and act in sector silos. This timely book illustrates the challenges of climate shocks to the resource nexus, assessing not just risk and vulnerability but the transmission of impacts through a system of systems. Dr. Howarth and co-authors draw on the Nexus Shocks project to demonstrate the importance of multi-sector collaboration and multi-agency co-production of resilience solutions.'
—Dr. Geoff Darch, *Anglian Water, UK*

Candice Howarth

Resilience to Climate Change

Communication, Collaboration and Co-production

Candice Howarth
University of Surrey
Guildford, UK

ISBN 978-3-319-94690-0   ISBN 978-3-319-94691-7   (eBook)
https://doi.org/10.1007/978-3-319-94691-7

Library of Congress Control Number: 2018947413

© The Editor(s) (if applicable) and The Author(s) 2019
This work is subject to copyright. All rights are solely and exclusively licensed by the Publisher, whether the whole or part of the material is concerned, specifically the rights of translation, reprinting, reuse of illustrations, recitation, broadcasting, reproduction on microfilms or in any other physical way, and transmission or information storage and retrieval, electronic adaptation, computer software, or by similar or dissimilar methodology now known or hereafter developed.
The use of general descriptive names, registered names, trademarks, service marks, etc. in this publication does not imply, even in the absence of a specific statement, that such names are exempt from the relevant protective laws and regulations and therefore free for general use. The publisher, the authors and the editors are safe to assume that the advice and information in this book are believed to be true and accurate at the date of publication. Neither the publisher nor the authors or the editors give a warranty, express or implied, with respect to the material contained herein or for any errors or omissions that may have been made. The publisher remains neutral with regard to jurisdictional claims in published maps and institutional affiliations.

Cover illustration: © nemesis2207/Fotolia.co.uk

This Palgrave Pivot imprint is published by the registered company Springer Nature Switzerland AG
The registered company address is: Gewerbestrasse 11, 6330 Cham, Switzerland

FOREWORD

In September 2015, 193 national governments signed up to the United Nations' seventeen Global Goals for sustainable development. Ensuring access for everyone to food, water and affordable, clean energy are at the forefront of this agenda, with a set of targets to be met by 2030. But the synergies, tensions and trade-offs between these goals are less well understood. Over the past decade, these have increasingly been analyzed and debated in terms of the 'nexus': a term which is not new, but has risen in prominence as a way of thinking about connections between food, water, energy and the wider environment.

Agreement on the Global Goals, followed weeks later by the Paris Agreement on climate change, increased optimism about the prospect of delivering on the 2030 sustainable development agenda. But in the three years since then, Britain's vote to leave the European Union, the election of a more protectionist American administration, and the growing strength of populist and authoritarian leaders in countries, such as Russia, Turkey and Hungary, have heightened concern about a slowing down or reversal of progress.

Such turbulence illustrates well the nexus argument: that efforts to improve the sustainability of one domain, without considering others, can easily fail, or create vulnerabilities and feedback loops of various kinds.

In this important and timely book, Candice Howarth sharpens our understanding of such interdependencies. Her notion of 'nexus shocks'—extreme weather events resulting from climate change, such as droughts, floods and heatwaves, which impact the nexus of food, water, energy and the environment—provides an illuminating heuristic through which to navigate the dilemmas and complexities of sustainability politics.

As Howarth reminds us, nexus shocks are likely to manifest themselves in diverse ways over the next 30 or 40 years—with different features and consequences for countries and communities such as those in the UK, Australia or Bangladesh, but one common challenge: how to develop resilient and integrated responses across multiple domains and levels of decision-making.

The conclusions that Howarth draws at the end of the book are considered and insightful, and her recommendations to policymakers, business and civil society, set out in clear and practical terms how to place collaboration, co-production and two-way communication at the heart of our preparedness for whatever nexus shocks may lie around the corner.

I am delighted that the Nexus Network, funded by the UK's Economic and Social Research Council (ESRC), and which I had the privilege to direct from 2014 to 2017, was able to support the research behind this book. The Nexus Network was set up to foster debate, support research and broker collaborations across food, energy, water and the environment.[1] Coordinated by the University of Sussex, with colleagues from the Universities of Cambridge, East Anglia, Exeter and Sheffield, the network engaged researchers from all disciplines, and decision-makers in government, business and civil society.

As the 2013 World Social Science Report put it: 'The social sciences must help to fundamentally reframe... global environmental change from a physical into a social problem.'[2] Social science has a particular contribution to make to this task. And social scientists—such as Candice Howarth—are at the forefront of developing interdisciplinary solutions to nexus challenges.

Working in these ways isn't easy. Academic reward systems still tend to privilege mono-disciplinary work. Career paths are less predictable and

---

[1]For more information on the Nexus Network, see: http://www.thenexusnetwork.org.

[2]http://www.worldsocialscience.org/activities/world-social-science-report/the-2013-report/.

more risky. More creative experiments—of which the Nexus Network is one example—will be required. And greater attention needs to be paid to the people who can make collaborations work; and the skills, training and capacity that they will need.

Some of these people will be researchers in universities; others will work in knowledge exchange, or in funding agencies. More will be found in the businesses, public bodies and NGOs that researchers need to partner with, if we are to succeed in scaling up the volume and intensity of collaborative, problem-oriented research and innovation—as this book advocates. All of these them are boundary spanners: the T-shaped people on which our capacity to navigate the nexus will ultimately rest.

Candice Howarth is one such boundary spanner. This book provides an invaluable resource and guide for those who want to join her.

James Wilsdon is Professor of Research Policy at the University of Sheffield and was Director of the ESRC Nexus Network from 2014 to 2017.

Sheffield, UK   James Wilsdon
                                                        University of Sheffield

ACKNOWLEDGEMENTS

The work presented and discussed in this book would not have been possible without the funding and support by the ESRC Nexus Network as part of a Nexus Networking Grant and the Nexus Shocks Fellowship. Thanks are extended to members of the Nexus Shocks Network, the London Climate Change Partnership, the Department of Sociology at the University of Surrey, the Global Sustainability Institute and the Nexus Network. In addition, further thanks go to Professor James Wilsdon, Professor Alison Anderson, Kristen Guida, Ian Christie and Geoff Darch for their invaluable feedback on the first draft of this book.

CONTENTS

About the Author and Co-authors

Candice Howarth is a Senior Lecturer in Sustainability and Climate Communication at the University of Surrey, UK. With a background in meteorology, climate change and behaviour change she is passionate about how the co-production of knowledge and climate communication can inform decision-making on sustainability challenges. She has led numerous projects on climate communication, the science-policy interface, nexus shocks and sustainable decision-making and has written extensively on these subjects. She is a member of the UCL Policy Commission on Communicating Climate Science, chairs the Nexus Shocks Network, sits on the Royal Meteorological Society Climate Science Communication group, and was a committee member of the British Sociological Association Climate Change Study Group. She sits on the Editorial Board of the journal *Environmental Communication*.

Katya Brooks is an independent researcher and former Research Fellow on the Nexus Shocks project at the University of Surrey. She has extensive experience of social research and project management with NGO, academia, government and business. With a background in sociology and human rights, she is particularly interested in social and environmental justice and participatory, reflexive and interdisciplinary approaches. Her recent work has focused on the community resilience, social vulnerability, community engagement and decision-making in relation to climate change. Katya contributed to the second phase of the Nexus Shocks project at the University of Surrey conducting an extensive evidence assessment. Dr. Brooks co-authored Chapter 1.

Sian Morse-Jones is a Senior Consultant at Collingwood Environmental Planning, specializing in social research in the environmental and sustainability field. Sian has contributed to and managed a range of projects with the private, public, NGO and academic sectors, in areas such as nexus shocks, natural capital accounting, ecosystem valuation and social impact assessment. Sian worked on the second phase of The Nexus Shocks Project at the University of Surrey undertaking qualitative research to explore stakeholder decision-making processes in relation to heatwaves and flooding with a focus on communication, collaboration and co-production. Dr. Morse-Jones co-authored Chapter 4.

LIST OF TABLES

CHAPTER 1

Introduction: Defining Nexus Shocks

Candice Howarth, Katya Brooks

Abstract This chapter introduces 'nexus shocks'. It explores who and what they impact and how, why they are important, and why the lens of nexus shocks provides a useful approach to practically explore and inform decision-making about climate shocks to food-energy-water-environment (FEWE) resources. Characteristics of nexus shocks are presented and discussed in the context of decision-making as well as how interpretation of these characteristics across stakeholder groups and sectors can lead to detrimental decision-making processes. The chapter closes with an overview of the Nexus Shocks project, the findings of which form the basis of this book.

Keywords Energy-food-water-environment nexus · Impacts
Responses · Stakeholders

HIGHLIGHTS

- Food, energy, water and environment resources and systems are deeply linked. The concept of 'nexus' can provide a useful lens through which to explore the interlinkages and interdependencies between them but its meanings, applications and implications are contested and varied in this context.

© The Author(s) 2019
C. Howarth, *Resilience to Climate Change*,
https://doi.org/10.1007/978-3-319-94691-7_1

- A nexus *shock* could be any sudden occurrence that disrupts and has knock-on effects across the food-energy-water-environment nexus and/or the actors that work within it. This book focuses specifically on extreme weather events or shocks resulting from climate variability and change.
- There is a need for an integrated approach at multiple levels across sectors and systems in preparing, planning, identifying, responding and making the most appropriate decisions to nexus shocks that incorporate the interlinkages between these core resources.
- A lack of national strategy to build resilience to nexus shocks can mean that climate-vulnerable places and people are not targeted and supported.

Understanding the Nexus

In sustainability research, 'nexus' refers to the intersection of the core resources: water, energy, food and environment, as a whole, interdependent and interlinked system. Since 2008, 'nexus' has gained in popularity as a tool to explore the relationships between whichever combination of these resources is under study (e.g. the food-energy-water-environment (FEWE) nexus, the water-energy-food (WEF) nexus), with the climate and of those working in science and environmental policy, governance and business spheres and across academic research disciplines (Leck et al. 2015; Rasul and Bikash 2016). This can also enable a better understanding of (un)intended consequences of policies, technologies and practices that may arise around nexus issues (Howarth and Monasterolo 2016).

The origin and use of 'the nexus' in this context can be traced back to the 2008 World Economic Forum and has since been used by former UK Chief Scientific Government Advisor Sir John Beddington, by prominent international institutions (e.g. the World Bank, the UN World Water Assessment Programme, the European Commission, the OECD), has been the focus of international events (e.g. Nexus Bonn 2011 Conference, Sixth World Water Forum in 2012, the Rio +20 negotiations in 2012, 2014 Stockholm Water Week) and has driven research priorities (e.g. UK Economic and Social Research Council (ESRC) Nexus Network). However, the term is not without its critics. Some argue that 'the nexus' is merely a buzzword (Cairns and Krzywoszynska 2016) to represent sustainability issues and it does not adequately consider local interactions between stakeholders and resources; resource inequality and access, and how this contributes to social instability; how decision-making varies across scales; nor, the role of science and technology in some of these debates (Howarth and Monasterolo 2017; Allouche et al. 2014). A single term can also not fully represent the key elements of the interactions and interdependencies that occur across food, energy, water, environment resources (Table 1.1). Terms such as the 'nexus' become labels of such complex processes that they become unable to fully represent the myriad of meanings, interpretations and interactions that occur with the nexus of food, energy, water, environment and indeed the real interactions that occur between those affected by or dependent on these resources. Whilst it provides a useful way of capturing such intricate processes, and is used widely in the academic sphere it is still of limited use in the political and industry environments (although it is a common term used by those working on issues relevant to nexus resources).

Nexus resources are fundamental for the evolution and development of society, however they are being depleted at a rate faster than ecosystems can cope with (Steffen et al. 2018). For example, between 2000 and 2010, global depletion of non-renewable groundwater increased by 24%, primarily through abstraction for agricultural irrigation (Dalin et al. 2017), and, due to consumption-based pressures, the planetary boundaries (Rockström et al. 2009) of nitrogen and phosphorus (two nutrients critical for plant and animal production, i.e. food systems), exceeded globally by more than factor of 3 and 2, respectively (Hoff et al. 2017). The concept of the nexus entails a

Table 1.1 Key elements that characterise the food-energy-water-environment nexus (adapted from Howarth and Monasterolo 2016)

Uncertainty about societal changes and political stances and uncertainty around how resources, stakeholders and processes are connected
Existence of **risks** which can lead to complex processes and can potentially lead to cascading or systemic effects
Impacts that can cascade across a system, amplified or diminished
Not always linear with numerous interacting components
Consisting of multiple **feedback** processes where a reaction from a component or event is enhanced (positive feedback) or diminished (negative feedback) following another component or event
Both **independent** of human intervention and **dependent** on physical and biophysical (e.g. environmental) mechanisms
Characterized by **hierarchy** with sub-systems all interacting with others

holistic view of the world that surrounds society and interactions with a complex system of feedback loops, different sectors and natural resources (Hamiche et al. 2016).

SHOCKS TO THE NEXUS

A nexus *shock* is anything or event that disrupts, changes or threatens the food, energy, water and environment nexus and/or the actors that work within this system and are heavily reliant upon these resources. In the context of this book, a nexus shock specifically refers to 'climate-related', 'severe weather' or 'extreme weather' events or shocks as direct effects of climate change, such as drought, floods and heatwaves. Globally, extreme weather and climate events have been changing with increased frequency and severity of such shocks observed since 1950 (IPCC 2013). A nexus shock can be characterized as a short-term event of low probability (i.e. not a high likelihood of occurring), low frequency (i.e. not occurring often), that has an immediate, high impact which affects a number of scales (i.e. local, national, regional, international) and stakeholders (i.e. businesses, policymakers, farmers, land owners, consumers, etc.) that work, contribute to, or depend on, the FEWE nexus (Howarth and Monasterolo 2016).

As part of an inherently interdependent, interlinked and complex system, a shock to one nexus resource can cause feedback loops for the others which can combine and cascade, and potentially compound vulnerabilities in each (Hamiche et al. 2016). These feedback loops mean one particular sector may heavily rely on one or more nexus resources (e.g. the food sector relying upon energy to transport produce and water to irrigate crops). Impacts on one of these may create negative or positive feedback loops (these could also be thought of as a vicious or virtuous circles) further down the chain (e.g. interruptions to energy or water supply will impact farm viability and food production which feedback into population growth). Thus, the nexus approach takes a holistic view of the world that surrounds society and interactions with a complex system of feedback loops, different sectors and natural resources. As such, it can enlighten the process where 'soft' factors such as human values and perceptions or the role of time in the interactions between different sectors influence how decisions are made. These can be challenging to monitor and measure but are crucial in delivering and supporting decision-making (Howarth and Monasterolo 2016). To minimize risks of impacts in future, nexus shocks require resilient responses to these impacts that ensure longevity, legacy and sustainability of decision-making processes.

The term 'risk' is often used instead of 'shock' in academic and practitioner discourse but their meanings and implications are slightly different (World Economic Forum 2018). The UN Development Programme's (UNDP) 2007 report *Climate Shocks and their Impact on Assets* employs the term 'climate shocks' and considers 'shocks as one step further in the definition of risks' whereby 'risks' are 'prospects of a shock', or a shock is 'the realization of a risk' (De La Fuente 2007) and Global Risk reports from 2007 to 2018 place food, energy and water shocks in the top five risks to the world in terms of likelihood and impact (World Economic Forum 2018). The word 'shock' is a better way of framing the topics explored in this book as it specifically suggests large size, low probability of occurrence and high likelihood of damage—particularly as people with high exposure and vulnerability and low resilience to shocks are likely to be those most affected by shocks and related stresses, such as physical or psychological strains (De La Fuente 2007). This distinction is useful as the phrase nexus 'shocks' helps represent events within a context of need for building resilience and capacities of human and natural society, systems or structures to cope with and adapt in order to respond better after the shock and in future.

Weather extremes resulting from climate change, are nexus shocks with impacts on social, economic, environmental and ecological development, international decision-making and the stability of financial markets. Understanding this can be challenging, particularly when considering how the context of climate change, a complex process in itself which alleviates or exacerbates climate impacts. When considering decision-making processes, these impacts can lead to delays in responses, short-term thinking, policy uncertainty and changes, unpredictability of impacts and can subsequently affect a number of outcomes (Jacobs and Mazzucato 2016; Hake et al. 2016). A key challenge is understanding the sources of uncertainties in our knowledge of these interactions, how they manifest themselves (e.g. through model projections or behavioural responses) and how they affect different scenarios for decision-making. Another challenge is the uncertainty of climate impacts on sectors within the nexus, such as how this could affect food production and food security at different geographical levels (Rosenzweig et al. 2014) and scales (Garcia and You 2016). Yet attempting to tackle these uncertainties may result in stakeholders and resources affected becoming even more vulnerable particularly if the impacts in question, how they affect these stakeholders and resources, and the exact nature of uncertainties are not fully understood. This may ultimately affect different sectors within the nexus with further social, economic, environmental and political implications. Tools that are used to model the dynamics of a complex (climate) system and how shock events can impact the nexus, have limitations themselves such as variability from one model to another, difficulty in estimating growth parameters of variables that feature in a model, time lags, inadequate representation of behavioural responses within individual sectors, and an underestimation of the negative effects of carbon and related social costs (Pindyck 2013; Farmer et al. 2015). The tools available to inform decision-making in responses to these shocks are therefore imperfect adding to the complexity and challenge of formulating responses that can align with the needs of all those across sectors affected by a nexus shock.

The Intergovernmental Panel on Climate Change's *Special Report on Managing the Risks of Extreme Events and Disasters* outlines exposures to climate shocks and the vulnerability of systems which are impacted by these shocks. It details how impacts may be exacerbated and affect resilience, adaptive capacity and the ability to cope immediately after a shock and in future (IPCC 2012). In the United Kingdom (UK), current government policy in relation to resilience and nexus shocks mainly centres on emergency planning and reactive decision-making processes. This focuses on building resilience to specific climate-related (e.g. floods) and non-climate-related (e.g. terrorist attacks) shocks. This also entails resilience of a wide range of entities such as communities, institutions and structures in the short-term (Johnson and McGuinness 2016). Emergency events of this nature are often responded to in a reactive way in the wider context of climate change which calls for a move to proactive decision-making, in order to adapt and transform to future nexus shocks, and build long-term resilience. The UK is affected by a number of climate-related nexus shocks as outlined in the UK Climate Change Risk Assessment (CCRA) (HM Government 2016); of these, flooding and heatwaves are ranked as the greatest direct threats.

Flooding, resulting from increases in heavy rainfall and associated risks of fluvial and surface flooding, is the most prominent nexus shock faced by communities in the UK (Pitt 2007; Committee on Climate Change 2016). If current levels of adaptation continue, and assuming a 2 °C increase in global temperatures, annual associated damages are expected to rise by 50% (Sayers et al. 2015) with costs of flooding reaching £10 billion in the next 25 years (Johnson and McGuinness 2016). This would lead to increased risk of infrastructure failures in energy, transport, water and communication. Longer-term stresses across the nexus will also be generated, including negative impacts on food and energy security, or pollution of agricultural land and food crops leading to rising costs of living (Preston et al. 2014). The UK Committee on Climate Change (2016) warns that the impacts of flooding and coastal change to communities, businesses and infrastructure are already at a significant level with evidence of this from UK floods experienced in Boscastle (2004), Carlisle (2005), Hull (2007), Cumbria (2009), and the UK winter floods of 2013–2014 and 2015 (Forrest et al. 2017). The shift from flood resistance to resilience in policy, attitude, roles and responsibilities of those involved in flood risk management responses and decision-making stimulated by these flood events is reflected in the UK government's national response plans to flooding events (Forrest et al. 2017): Making Space for Water strategy (Defra 2005), the Pitt Review (Pitt 2007), the Flood and Management Act (HM Government 2010), UK CCRA 2017: Projections of Future Flood Risk (Sayers et al. 2015), and the National Flood Resilience Review (HM Government 2016). In 2016, the Government established Flood Re to make affordable flood insurance available to households in flood risk areas and led to an initiative to build resilience to flood risk at the local level (Flood Resilience Community Pathfinder scheme) in 2013–2015. Supporting 13 projects led by local authorities in areas of significant or greater risk of flooding across England, the scheme aimed to assess factors that help build resilience, support the implementation of physical interventions (e.g. property level protection) and encourage new approaches to partnership working, collaboration and communication, particularly between local authorities and communities (Defra 2012). The projects led to the establishment and maintenance of 111 community flood groups, forums and networks (Twigger-Ross et al. 2015). Setting these up through multi-agency partnerships involving other local authority departments, national agencies (e.g. Environment Agency), civil society and communities proved invaluable to developing these institutional structures, governance processes and networks which have been important for building institutional resilience and linking with the wider resilience agenda (Twigger-Ross et al. 2015).

A second important climate risk identified by the UK CCRA is heatwaves. There is currently no universal definition of heatwaves (Benzie et al. 2011) and this has important implications for communication, decision-making and shared understanding (Howarth and Brooks 2017). The UK Meteorological (Met) Office (2018) uses the World Meteorological Organization's (WMO) definition: '...when the daily maximum temperature of more than five consecutive days exceeds the average maximum temperature by 5 degrees Celsius, the normal period being 1961–1990' (where the average maximum temperature refers to the average for a specific location and time of year) and Guerreiro et al. (2018) define heatwaves as 'three consecutive days where both the maximum and the minimum temperature exceed their respective 95th percentile from the historical period' (p. 2). Heatwaves are currently rare in the UK but it is predicted that future climate and demographic changes combined with urban development will increase the frequency, intensity and duration of heatwaves (Smith et al. 2016). Public Health England (PHE) published its UK Heatwave Plan as a result of the devastating heatwave experienced across Europe in the summer 2003. The shock of a severe and prolonged heatwave such as this can negatively impact food, water and energy supplies as well as businesses, transport, health and social care services with effects felt at the local level. The CCRA 2017 predicts summer heatwaves like those experienced in 2003 are expected to become the norm by the 2040s and premature heat-related deaths to more than triple by the 2050s (Committee on Climate Change 2016). PHE's plan describes the five-tiered (Levels 0–4) Heat-Health Watch system which operates over the summer months in England from 1 June to 15 September, supported by heatwave alerts co-produced with the Met Office. The factors that make people, systems and structures vulnerable to high temperatures are complex and dynamic, and include quality of housing and the built environment, local urban geography, household income, employment, tenure, social networks and self-perception of risk (Benzie 2014). Similarly to risks of flooding, these factors affect people's ability to respond in order to adapt to these extreme heat conditions. For example, usually during high temperatures in the UK heat risks tend not to be perceived as personal risks and people seek to spend more time outdoors with little consideration of the health impacts. To address this issue, PHE, the Met Office, National Health Service (NHS), Age UK and Cancer Research work collaboratively to co-produce campaigns communicating health advice during heatwaves, particularly for vulnerable groups.

Flooding and heatwaves are characterized by multiple meteorological, environmental and geographic complexities and impact society and the nexus in a number of ways, both directly and indirectly. Responses to these shock events have often been driven by a severe historical shock (e.g. the 2013–2014 UK winter floods or the 2003 European heatwave) and have involved leadership, communication and collaboration across sectors, scales and stakeholders, with scientific advice underpinning much of this process (Howarth and Painter 2016). Heatwaves and flooding require collaboration between a number of stakeholders at a range of scales with different (and occasionally conflicting) priorities: public bodies such as local authorities and Local Resilience Forums; private organisations such as utility companies and internal drainage boards; and civil society organisations such as the National Flood Forum and community groups. However, when these shocks occur, due to their 'shock-like' nature, decision-making may not 'follow the norms of rationality' (Castan Broto and Bulkeley 2013) and information provided to inform that decision-making might not necessarily meet the needs of those using it (White and Stirling 2012).

DECISION-MAKING AND NEXUS SHOCKS

The concept of *nexus thinking* sits well with innovative approaches aimed at better understanding characteristics and interactions that would enable decision-makers to better address sustainability challenges such as responses to nexus shocks. However, more knowledge is needed on how stakeholders from academia, policy, business, finance and insurance sectors perceive and experience the nexus and impacts of shocks to it. Similarly, an understanding of the dynamics of their network of relationships within the nexus and the potential effects of other actors working across the nexus on decision-making processes is needed to increase resilience to climate shocks. An integrated and transdisciplinary approach will help increase understanding of how to navigate the complexities within human-environmental systems to develop effective solutions and decision-making processes (Howarth and Monasterolo 2017).

A linear, single-discipline process of decision-making is often assumed among decision-makers. However, in proactive preparation and adaptation before and during times of shocks, multiple areas of governance spanning different sectors of expertise across the FEWE nexus are required to coordinate and resolve urgent issues and jointly identify the most appropriate responses. In some cases there could be greater clarity between preparation and responses to shocks that we experience now, and preparation for shocks that we expect to be more significant in future. Reframing the nature of decision-making as a process of aligning multiple actors and strategies opens opportunities to examine issues within decision-making processes. The nexus displays a number of complexities, opportunities and challenges when it comes to informing decision-making in response to nexus shocks such as heatwaves and flooding, which are often interdisciplinary, cross-cutting and multi-sectorial. These complexities are not solely limited to one sector, which are linked due to the dependencies between food, energy and water resources. Consequently, interactions and shocks to one of these resources will inevitably impact one or more other sector. The complex, non-linear, space and time dependency of the shocks means a shock may originate in one area of the nexus and trickle down onto others. There is a need to move away from current sector-based approaches (that focus on solutions led by or involving single sectors) to knowledge development and solution creation that span multiple disciplines and sectors.

Important decisions on how to increase resilience to nexus shocks are based on sound scientific advice yet in a number of cases, this advice is based on multiple scenarios (e.g. based on different model projections which result from different models using the same data or using different assumptions to underpin their model runs) where there can be large uncertainty in the projected outcomes. Decision makers therefore need to plan for a wide range of possible outcomes and assess which is most probable in terms of its magnitude and likelihood of occurring. One such example is the London Thames Barrier and the Adaptive Pathways that were developed through collaboration between the UK's Department on Energy, Food and Rural Affairs (Defra), the Environment Agency and the Met Office. The potential shocks of increased flooding led to a debate on whether and when to build a new Thames Barrier to protect the City of London (Lowe et al. 2009) from severe flooding. Doing so too early (i.e. 20 years too soon) would cost a significant amount when there was no imminent threat and so the need for Adaptive Pathways enabled a range of options to be constructed based on the scientific advice and scenarios available, providing the ability for flexibility where switching from one potential scenario to another depending on 'real world' need. Flooding and heatwaves are identified with 'high confidence' by the IPCC (2014) as key future risks caused by climate change that span sectors and regions worldwide. Hence the next section explores international examples from Bangladesh and Australia of the shocks and impacts across nexus resources and the implicated actors, responses and decision-making processes.

HEATWAVES IN AUSTRALIA

Evidence from across Australia of the global trend towards increasing frequency and severity of nexus shocks is stark: 2017 was the third hottest year since records began in 1910 (CSIRO and Bureau of Meteorology 2017) and between December 2016 and February 2017 more than 200 climate records were broken (Steffen et al. 2017). The national average temperature has increased by 0.9 °C since records began in 1910 and without radical action, temperatures are projected to rise by 3–5 °C in coastal areas and 4–6 °C inland by 2100 (Bureau of Meteorology 2016; IPCC 2014; Climate Change Authority 2015; Lewis and King 2015). Heatwaves have beset much of the country every year since 2013 (NOAA 2018) and have taken more lives than any other natural hazard since 1890 (Australian Government 2016). Since 1950, the annual number of record hot days across Australia has more than doubled and 2017 was Australia's third hottest year since 1910 (CSIRO and Bureau of Meteorology 2017). Between December 2016 and February 2017 more than 200 climate records were broken across the country, including: the hottest summer on record for Brisbane and Sydney with temperatures topping 47 °C and the town of Moree experienced 54 consecutive days of 35 °C or above (Steffen et al. 2017).

The direct impacts of this record-breaking heat and its interactions with other climate shocks and environmental stresses, particularly drought and bushfire, are cumulative and widespread across the FEWE nexus, human health, and physical and social infrastructures. Heatwaves can also result in significant economic costs from lost productivity resulting from energy and transport disruptions, sick leave compensation claims due to heat stress (Xiang et al. 2015) and damage to crops, infrastructure and property. For example, the Australian winter of 2017 was the hottest and driest on record; this led to an extended and exacerbated bushfire season across much of New South Wales (NSW) that destroyed at least 32 homes and caused an estimated AU $20 million of property damage (Insurance Council of Australia quoted in Steffen et al. 2017). In response, the NSW State Government's Department for Justice offered disaster assistance to help communities with the cost of recovery through the jointly-funded Commonwealth-State Natural Disaster Relief and Recovery Arrangements (NDRRA) (New South Wales Government 2017).

A national survey conducted in 2015 found high rates of non-insurance and under-insurance across Australia: 13% of respondents were without insurance cover for assets, 9% were without home insurance, and 41% of tenants did not have contents insurance (Booth et al. 2015). A report commissioned by Lloyds estimated the average uninsured loss for each Australian natural disaster between 2004 and 2011 as almost AU $1 billion (Edwards and Davis 2012). Insurance and subsequent supporting systems can help to reduce the impacts of nexus shocks and aid recovery, adaptation and transformation but only in conjunction with other interventions from all levels and sectors, such as government assistance, collective clean-ups and infrastructure repairs (Booth et al. 2015). The February 2017 heatwave also left 7000 properties without electricity in New South Wales as energy systems failed and a major traffic jam on the Hume Highway between Sydney and Melbourne was caused by asphalt melting whilst railway tracks buckled in the heat (Steffen et al. 2017: 14).

Australia has no national heatwave plan but in 2011, PriceWaterhouseCooper (PwC) in collaboration with the Australian Government published a national framework to advise on how to better respond to heatwaves. This prompted development of the Heatwave Service forecasting tool and the first official Australian definition of a heatwave by the Bureau of Meteorology (2016): 'three days of high maximum and minimum temperatures that is unusual for that location'. Like the UK Met Office's heat-health-watch system, the Australian Heatwave Service operates through the summer months, classifies heatwaves relatively by using local average temperatures and predicted health impacts, and provides a set of maps showing heatwave severity. The Heatwave Service is not yet fully integrated with the Bureau's 'other analysis, forecast and warning services meaning there will not be a guaranteed notification of heatwaves with other forecast services, and the service may not be available in the event of a system malfunction' (Bureau of Meteorology 2018). It is also not yet possible for vulnerable persons to receive a personal heatwave warning unless they have internet access. The incidence of severe and prolonged heatwaves in Australia, and worldwide, is projected to increase

(PwC 2011) and such limitations in early warning systems pose serious consequences across communities and the nexus. To reduce impacts, it is crucial that actors from all sectors are well prepared and coordinated with a collaborative decision-making approach all levels: a national heatwave response plan that considers impacts across the nexus, an integrated, resilient forecasting service and an overarching policy framework that builds on what has already been achieved (PwC 2011; Steffen et al. 2014).

EXTREME FLOODING IN BANGLADESH

Due to its geographic and climatic features, Bangladesh is one of the most climate-vulnerable and flood-prone countries in the world, especially during the monsoon season, and the scale, duration and intensity of floods to the region are increasing (WMO 2017). In August 2017, a humanitarian crisis unfolded across South Asia as monsoon floods affected over 40 million people and destroyed thousands of homes, schools, hospitals and farms (Humanitarian Coordination Task Team (HCTT) 2017). From March 2017, Bangladesh experienced four significant floods and the fourth, in August, was the most severe in 100 years: 32 of the country's 64 districts were submerged, over 8 million people were affected and at least 145 lives were lost (HCTT 2017). The impacts of the floods across the food, energy, water, environment nexus included: contamination of over 65,000 tube wells; damage or destruction of over 700,000 houses, 11,000 km roads, 500 bridges and culverts, 1000 km railway lines; and, losses of livestock, fish stocks and over 650,000 hectares of standing crops (FAO 2017; HCTT 2017). Longer-term stresses generated across the nexus link to priority concerns for flood-affected communities identified by the Bangladesh Needs Assessment Working Group's (NAWG) 72-hour assessment: livelihood, shelter and food security (HCTT 2017). The impact of floods on agriculture has a role in each of these. With rice among the worst affected crops and 1.5 million people in need of food assistance, rice price spikes reached record levels in September 2017 (FAO 2017; HCTT 2017). The Government was forced to import 1.5 million tons of rice after six years of self-sufficiency (Chaudhary and Pradhan 2017). The agriculture sector contributes about 16% of the country's GDP and employs approximately 45% of the population; as a result, flooding has compromised livelihood security (FAO 2017; HCTT 2017), combined with rises in living costs, the means to rebuild homes (shelter) is severely limited.

Similarly to the UK, a strategic shift from flood resistance to a more participatory, resilience approach in response and decision-making for flood events is evident with a range of programmes implemented by the Bangladesh Government: the National Adaptation Programme of Action (2005); Disaster Risk Reduction and Emergency Response Management Plan (2012); Flood Response Preparedness Plan (2013), National Resilience Programme (2017). This national strategy to build community resilience to nexus shocks makes it more likely that areas of climate vulnerability are prioritized and targeted. Limited data are available to fully evaluate the financial cost of these floods but according to the World Resources Institute's (2015) global flood analyzer an estimated US $5.4 billion of the country's GDP is at risk from the impacts of floods annually. Bangladesh received US $27,668,866 in international humanitarian funding in response to the 2017 flooding emergency (UNOCHA 2018). Following severe flooding in 2014 a multi-stakeholder partnership was established including Oxfam and insurance companies Pragati and Swiss Re to deliver an index-based flood insurance scheme to support communities in Bangladesh. In the pilot, 708 families in the Sirajganj district each received US $35 (2800 Bangladeshi taka) triggered by a policy that started in 2013 (Oxfam 2015; Swiss Re 2015). Decision-making in response to nexus shocks in Bangladesh involves international, national, local, community and non-governmental organizations and mechanisms. The processes, systems and capacities needed at all levels to respond, recover, adapt and transform from flood shocks in Bangladesh are complicated by underlying economic development, social and climate disadvantage, inequality and vulnerability issues. Nexus shocks are increasingly the limiting factor for economic development and future security of food, energy and water, therefore, it is vital to continue to ask questions about power, politics and why vulnerability exists (Mohtar and Daher 2016). There is a role for the government to develop a strategic collaborative approach between community and top-down interventions to help ensure long-term resilience to flooding and other nexus shocks and focus on climate vulnerable areas.

The Nexus Shock Project

This book draws on data produced from an innovative programme of transdisciplinary research which brought together over 300 stakeholders from academic, business, government and not-for-profit organisations in the UK, all interested in decision-making processes in the context of resilience to nexus shocks. The Nexus Shocks project sought to improve decision-making processes and resilience related to nexus shocks and was funded in 2015 by the UK's ESRC's Nexus Network. It adopted a co-production approach and brought together interdisciplinary and cross-sectorial expertise to engage in constructive dialogue, identify opportunities to address challenges, and explore how to better inform decision-making in response to these shocks. The aims of this work, were threefold: firstly to overcome obstacles and build on opportunities in responses to nexus challenges, secondly to innovatively assess the complexity of societal responses to nexus shocks, and thirdly to better inform business and policy responses. This book presents important and impactful findings of the two consecutive phases of the Nexus Shocks project.

Phase 1 of the project ran from April to December 2015. Five workshops were conducted in collaboration with the Met Office, Atkins, Chatham House, Lloyds of London, Wills Re, Climate UK, Cambridge Cleantech and LDA Design. The workshops aimed to explore how climate and weather shocks affect the energy-food-water-environment nexus and focused on the following themes: Predicting shocks and hazards; Transmission and mitigation of nexus risks through infrastructure; Insurance and finance for resilience; Local business responses to shocks; and Governance, governments and shocks. Phase 2, the Nexus Shocks Fellowship, ran from October 2016 until September 2017. It built on work from Phase 1 with a view to improving responsiveness in shock scenarios affecting the FEWE nexus. The phase comprised a desk-based review of the literature (Howarth and Brooks 2017) and a qualitative study to collect primary data using thirty semi-structured interviews with individuals from academia, policy and practitioner communities in the UK. It explored insights from existing evidence and practice on the current picture of decision-making and resilience to nexus shocks in the UK and how this could be enhanced in different contexts.

To improve effectiveness of those working at the science-policy interface—including academics, policy makers and practitioners—the Nexus Shocks project explored some of the practical challenges associated with the integrated and transdisciplinary approach required to effectively manage and respond to nexus shocks. With the likelihood of such shocks predicted to increase in the future due to a changing climate, the need for effective, integrated decision-making across sectors, and including a number of stakeholders, will become increasingly important. The nature of decision-making involves multiple people, organizations, sectors and strategies, opens opportunities to examine issues within decision-making processes that occur during nexus shocks and identifies ways to improve resilience to shocks. Given the uncertainty surrounding information on nexus shocks, governments are faced with making challenging decisions based on uncertain science. To capture the above complexities, encourage collaboration across disciplines, and ensure engagement with appropriate stakeholders, the work presented in the book is truly unique. It adopted a co-production approach to its design, where all stakeholders involved in decision-making in response to nexus shocks (and those affected by it) were invited to take part in designing the project. In doing so, the findings shed light on important insights from scientific, policy and practitioner knowledge on nexus shocks.

References

Allouche, J., Middleton, C., & Gyawali, D. (2014). *Nexus nirvana or nexus nullity? A dynamic approach to security and sustainability in the water-energy-food nexus* (STEPS Working Paper 63). Brighton, UK: STEPS Centre.

Australian Government. (2016). *State of the environment 2016*. Canberra, Australia: Australian Government. Accessed: https://soe.environment.gov.au/theme/overview.

Benzie, M. (2014). Social justice and adaptation in the UK. *Ecology and Society, 19*(1), 39.

Benzie, M., Harvey, A., Burningham, K., Hodgson, N., & Siddiqi, A. (2011). *Vulnerability to heatwaves and drought: Case studies of adaptation to climate change in South-West England*. York, UK: Joseph Rowntree Foundation.

Booth, K., Tranter, B., & Eriksen, C. (2015). Properties under fire: Why so many Australians are inadequately insured against disaster. *The Conversation*. Accessed: https://theconversation.com/properties-under-fire-why-so-many-australians-are-inadequately-insured-against-disaster-50588.

Bureau of Meteorology. (2016). *State of the climate 2016*. Canberra: Australian Government Bureau of Meteorology. Accessed: http://www.bom.gov.au/state-of-the-climate/.

Bureau of Meteorology. (2018). *About the Heatwave Service*. Australian Government Bureau of Meteorology. Available online at http://www.bom.gov.au/australia/heatwave/about.shtml.

Cairns, R., & Krzywoszynska, A. (2016). Anatomy of a buzzword: The emergence of 'the water-energy-food nexus' in UK natural resource debates. *Environmental Science & Policy, 64,* 164–170.

Castan Broto, V., & Bulkeley, H. (2013). A survey of urban climate change experiments in 100 cities. *Global Environmental Change: Human Policy Dimensions, 23,* 92–102.

Chaudhary, A., & Pradhan, B. (2017). Floods may cost South Asia $215 billion a year by 2030. *Bloomberg online.* Accessed: https://www.bloomberg.com/news/articles/2017-09-12/south-asia-cities-face-215-billion-worth-extreme-rainfall-risks,13/9/17.

Climate Change Authority. (2015, July 2). *Final report of Australia's future emissions reduction targets.* Canberra: Australian Government Climate Change Authority.

Committee on Climate Change. (2016). *UK climate change risk assessment 2017 synthesis report: Priorities for the next five years.* Accessed: https://www.theccc.org.uk/wp-content/uploads/2016/07/UK-CCRA-2017-Synthesis-Report-Committee-on-Climate-Change.pdf.

CSIRO and Bureau of Meteorology. (2017). *Climate change in Australia: Technical report 2007.* Canberra: CSIRO and Bureau of Meteorology.

Dalin, C., Wada, Y., Kastner, T., & Puma, M. (2017). Groundwater depletion embedded in international food trade. *Nature, 543,* 700–704.

Defra. (2005). *Making space for water—Taking forward a new government strategy for flood and coastal erosion risk management in England.* London, UK: Department for Environment, Food and Rural Affairs.

Defra. (2012). *Flood risk community pathfinder scheme prospectus.* London, UK: Department of Food, Rural Affairs and Agriculture.

De La Fuente, A. (2007). *A climate shocks and their impact on assets.* New York: United Nations Development Programme.

Edwards, C., & Davis, C. (2012). *Lloyd's global underinsurance report.* London, UK: The Society of Lloyd's.

FAO. (2017, October 3). *Global Information and Early Warning System on Food and Agriculture (GIEWS) update: Bangladesh.* Rome: Food and Agriculture Organization of the United Nations.

Farmer, J., Hepburn, C., Mealy, P., & Teytelboym, A. (2015). A third wave in the economics of climate change. *Environmental & Resource Economics, 62*(2), 329–357.

Forrest, S., Trell, E.-M., & Woltjer, J. (2017). Flood groups in England: Governance arrangements and contribution to flood resilience. In E.-M. Trell, B. Restemeyer, M. Bakema, & B. Van Hoven (Eds.), *Governing for resilience in vulnerable places.* London, UK: Routledge.

Garcia, D. J., & You, F. (2016). The water-energy-food nexus and process systems engineering: A new focus. *Computers & Chemical Engineering, 91,* 49–67.

Guerreiro, S. B., Dawson, R. J., Kilsby, C., Lewis, E., & Ford, A. (2018). Future heat-waves, droughts and floods in 571 European cities. *Environmental Research Letters, 13*(3). https://doi.org/10.1088/1748-9326/aaaad3.

Hake, J.-F., Schlor, H., Schurmann, K., & Venghaus, S. (2016). Ethics, sustainability and the water, energy, food nexus approach—A new integrated assessment of urban systems. *Energy Procedia, 88*, 236–242.

Hamiche, A. M., Stambouli, A. B., Flazi, S. (2016). A review of the water-energy nexus. *Renewable and Sustainable Energy Reviews, 65*, 319–331.

HCTT. (2017). *Bangladesh Humanitarian Coordination Task Team (HCTT)— Situation report number 5: Monsoon floods in Bangladesh.* Accessed: https://reliefweb.int/sites/reliefweb.int/files/resources/HCTTSitrep_Monson%20 Floods_Bangladesh_Final.pdf.

HM Government. (2010). *Flood and water management act.* London, UK: HM Government.

HM Government. (2016). *National flood resilience review.* London, UK: HM Government.

Hoff, H., Häyhä, T., Cornell, S., & Lucas, P. (2017). *Bringing EU policy into line with the planetary boundaries.* Stockholm: Stockholm Environment Institute.

Howarth, C., & Brooks, K. (2017). Decision-making and building resilience to nexus shocks locally: Exploring flooding and heatwaves in the UK. *Sustainability, 9*, 838–854.

Howarth, C., & Monasterolo, I. (2016). Understanding barriers to decision-making in the UK energy-food-water nexus: The added value of interdisciplinary approaches. *Environmental Science & Policy, 61*, 53–60.

Howarth, C., & Monasterolo, I. (2017). Opportunities for knowledge co-production across the energy-food-water nexus: Making interdisciplinary approaches work for better climate decision-making. *Environmental Science & Policy, 75*, 103–110.

Howarth, C., & Painter, J. (2016). Exploring the science-policy interface on climate change: The role of the IPCC in informing local decision-making in the UK. *Palgrave Community, 2*, 16058.

IPCC. (2012). Summary for policymakers. In C. B. Field, V. Barros, T. F. Stocker, D. Qin, D. J. Dokken, K. L. Ebi, et al. (Eds.), *Managing the risks of extreme events and disasters to advance climate change adaptation* (A special report of working groups I and II of the intergovernmental panel on climate change). Cambridge and New York: Cambridge University Press.

IPCC. (2013). Summary for policymakers. In T. F. Stocker, D. Qin, G.-K. Plattner, M. Tignor, S. K. Allen, J. Boschung, et al. (Eds.), *Climate change 2013: The physical science basis. Contribution of working group I to the fifth assessment report of the intergovernmental panel on climate change.* Cambridge and New York: Cambridge University Press.

IPCC. (2014). *Climate change 2014: Synthesis report. Contribution of working groups I, II and III to the fifth assessment report of the intergovernment panel on climate change* [Core Writing Team: R. K. Pachauri and L. A. Meyer (Eds.)]. Geneva: IPCC.

Jacobs, M., & Mazzucato, M. (2016). *Rethinking capitalism: Economics and policy for sustainable and inclusive growth*. Chichester, UK: Wiley-Blackwell.

Johnson, N., & McGuinness, M. (2016, October 17–21). *Flood resilience in the context of shifting patterns of risk, complexity and governance: An exploratory case study*. E3S FLOODrisk 2016, 3rd European Conference on Flood Risk Management, Lyon, France.

Leck, H., Conway, D., Bradshaw, M., & Rees, J. (2015). Tracing the water-energy-food nexus: Description, theory and practice. *Geography Compass, 9*(8), 445–460.

Lewis, S., & King, A. (2015). Dramatically increased rate of observed hot record breaking in recent Australian temperatures. *Geophysical Research Letters, 42*(18), 7776–7784.

Lowe, J., Howard, T., Pardaens, A., Tinker, J., Jenkins, G., Ridley, J., et al. (2009). Thames Estuary 2100 case study. In *UK climate projections science report: Marine and coastal projections* (Chapter 7). Exeter, UK: UK Climate Projections.

Met Office. (2018). *What is a heatwave?* Accessed: http://www.metoffice.gov.uk/learning/learn-about-the-weather/weather-phenomena/heatwave.

Mohtar, R., & Daher, B. (2016). Water-energy-food nexus framework for facilitating multi-stakeholder dialogue. *Water International, 41*(5), 655–661.

Mora, C., Dousset, B., Caldwell, I., Powell, F., Geronimo, R., Bielecki, C., et al. (2017). Global risk of deadly heat. *Nature Climate Change, 7*, 501–506.

New South Wales Government. (2017, May 4). *Disaster assistance for bushfire affected communities in NSW*. Sydney: Department of Justice, Office of Emergency Management Online. Accessed: https://www.emergency.nsw.gov.au/Pages/media-releases/2017/Disaster-assistance-for-bushfire-affected-communities-in-NSW-May-2017.aspx.

NOAA. (2018). *Global analysis—Global climate report—Annual 2017*. Accessed: https://www.ncdc.noaa.gov/sotc/global/201713.

Oxfam. (2015, February 25). *Insuring against flood risk in Bangladesh*. Oxford: Oxfam. Accessed: wee.oxfam.org/profiles/blogs/insuring-against-flood-risk-in-Bangladesh.

Pindyck, R. (2013). Climate change policy: What do the models tell us? *Journal of Economic Literature, 51*(3), 860–872.

Pitt, M. (2007). *Learning lessons from the 2007 floods: An independent review*. London, UK: Cabinet Office.

Preston, I., Banks, N., Hargreaves, K., Kazmierczak, A., Lucas, K., Mayne, R., et al. (2014). *Climate change and social justice: An evidence review*. New York, UK: Joseph Rowntree Foundation.

PwC. (2011). *Protecting human health and safety during severe and extreme heat events: A national framework*. Canberra: PriceWaterhouseCooper and Australian Government. Accessed: https://www.pwc.com.au/industry/government/assets/extreme-heat-events-nov11.pdf.

Rasul, G., & Bikash, S. (2016). The nexus approach to water–energy–food security: An option for adaptation to climate change. *Climate Policy, 16*(6), 682–702.

Rockström, J., Steffen, W., Noone, K., Persson, Å., & Chapin, F. S., III. (2009). Planetary boundaries: Exploring the safe operating space for humanity. *Ecology & Society, 14*(2), 32.

Rosenzweig, C., Elliott, J., Deryng, D., Ruane, A., Müller, C., Arneth, A., et al. (2014). Assessing agricultural risks of climate change in the 21st century in a global gridded crop model inter-comparison. *Proceedings of the National Academy of Sciences of the United States of America, 111*(9), 3268–3273.

Sayers, P., Horritt, M., Penning-Rowsell, E., & McKenzie, A. (2015). *UK Climate change risk assessment 2017: Projections of future flood risk*. London, UK: Committee on Climate Change.

Smith, S., Elliot, A., Hajat, S., Bone, A., Smith, G., & Kovats, S. (2016). Estimating the burden of heat illness in England during the 2013 Summer heatwave using syndromic surveillance. *Journal of Epidemiology Community Health, 1*, 1–7.

Steffen, W., Hughes, L., & Perkins, S. (2014). *Heatwaves: Hotter, longer, more often*. Climate Council of Australia. Accessed: https://www.climatecouncil.org.au/uploads/9901f6614a2cac7b2b888f55b4dff9cc.pdf.

Steffen, W., Rice, M., & Alexander, D. (2017). *Cranking up the intensity*. Climate Council of Australia. Accessed: https://www.climatecouncil.org.au/uploads/1b331044fb03fd0997c4a4946705606b.pdf.

Steffen, W., Rice, M., & Alexander, D. (2018). *2017: Another record-breaking year for heat and extreme weather*. Climate Council of Australia. Accessed: http://www.climatecouncil.org.au/uploads/8e9c2b91ce3c3ebb-7d97e403a6fdf38e.pdf.

Swiss Re. (2015). *Closing the gap: Flood protection for Bangladesh*. Zurich: Swiss Re Insurance. Accessed: media.swissre.com/documents/Swiss+Re+case+study_Closing+the+gap_Bangladesh_FINAL.pdf.

Twigger-Ross, C., Orr, P., Brooks, K., Sadauskis, R., Deeming, H., Fielding, J., et al. (2015). *Flood resilience community pathfinder evaluation final report*. London, UK: Defra.

UNOCHA. (2018). *Financial tracking service*. Accessed: https://fts.unocha.org.

WEF. (2018). *The global risks report 2018* (13th ed.). Geneva: World Economic Forum.

White, R., & Stirling, A. (2012). Sustaining trajectories towards sustainability: Dynamics and diversity in UK communal growing activities. *Global Environmental Change, 23,* 838–846.

WMO. (2017). Statement on the state of global climate in 2017—Provisional release. Geneva: World Meteorological Association. Accessed: https://public.wmo.int/en/media/press-release/2017-set-be-top-three-hottest-years-record-breaking-extreme-weather.

Xiang, J., Hansen, A., Pisaniello, D., & Peng, B. (2015). Extreme heat and occupational heat illnesses in South Australia, 2001–2010. *Occupational and Environmental Health, 78*(8), 580–586.

CHAPTER 2

Mitigating and Exacerbating Climate Shocks to the Nexus

Abstract This chapter explores how responses to nexus shocks can help reduce impacts or make them worst. It draws on findings from five co-production workshops with the UK Met Office, Atkins, Chatham House, Lloyds of London and Willis Re, Cambridge Cleantech and LDA Design, to assess the factors that exacerbate and mitigate climate shocks to the food, energy, water, environment nexus and subsequent impacts. These are especially important to consider, as they enable opportunities for lessons learnt and better and more resilient responses to nexus shocks in future. However, these are often inadequately explored, and more is needed to ensure decision-making remains relevant and aligned with the needs of stakeholders affected by nexus shocks, when dealing with the complex nature of these shocks.

Keywords Decision-making · Climate shock · Mitigating
Exacerbating · Communication · Resilience

HIGHLIGHTS

- Responses to climate shocks can make initial impacts across the energy, food, water, environment sectors worst if inadequately managed, however, these factors can also help mitigate future shocks.

© The Author(s) 2019
C. Howarth, *Resilience to Climate Change*,
https://doi.org/10.1007/978-3-319-94691-7_2

- Exacerbators of shocks, when properly understood, identified and implications examined, may not be fully or permanently removed, and can become obstacles that exist within the system.
- Impacts of climate shocks to the food, energy, water, environment nexus demonstrate clear similarities in terms of the prediction of shocks, interactions with infrastructure, local shock experiences, the role of finance and insurance, and governments and governance, when explored through an exacerbation/mitigation lens.

Nexus shocks are growing in prominence on the global stage. All five environmental risks within the World Economic Forum's (WEF) 2018 Global Risk Perception Survey ranked higher than average for both likelihood of occurring and impact of the risk. This reflects the year 2017 which was characterized by increases in CO_2 prominent hurricanes, and extreme temperature records resulting in resources across the food, energy, water, environment nexus being under increasing strain and growing in vulnerability. Extreme weather events are classified by the WEF as the top risk in terms of likelihood and the second highest risk in terms of impact, whereas failure of climate change mitigation and adaptation features as the fifth highest likely risk and the fourth most impactful risk (WEF 2018). Yet, many of the top risks that featured in the Global Risk Report are connected in some way to climate change and weather extremes: large-scale involuntary migration, man-made environmental disasters feature as the sixth and seventh most likely risks respectively, and water crises, food crises, biodiversity loss and ecosystem collapse, large-scale involuntary migration and spread of infectious diseases are listed as the fifth, seventh, eighth, ninth and tenth risks, respectively, in terms of impact. This demonstrates an increased recognition of the prominence of these shocks with warnings against complacency as shocks can arise unexpectedly with profound impacts; '*in a world of complex and interconnected systems, feedback loops, threshold effects and cascading disruptions can lead to sudden and dramatic breakdowns*' (WEF 2018: 7).

Assessing the impacts from climate shocks to assets within the food, energy, water, environment nexus depends on the sensitivity (i.e. the extent to which an asset can survive the stress imposed upon it by the shock) and resilience (i.e. 'the ability of a system, entity, community or person to adapt to a variety of changing conditions and to withstand shocks while still maintaining its essential functions', World Bank 2015) of these assets (De la Fuente 2007). With a growing global population, predicted to exceed 9.7 billion by 2050 (World Bank 2017), this will lead to increasing pressure on already overloaded systems and depleting water and food resources. Resilience is crucial to ensure adequate responses to climate shocks. Minimizing impacts across the nexus is needed to ensure the world's most vulnerable and poorest, who will be the worst impacted by these shocks, suffer minimal losses of homes and property and do not endure great health, mortality or morbidity impacts (World Bank 2015).

Decision-making processes to respond to and alleviate the impacts from nexus shocks are subject to a range of elements that make these impacts better or worst. Robust decision-making often assumes that it is 'better to be roughly right than precisely wrong' (Bhave et al. 2016: 3), this is also true in the case of nexus shocks where immediate responses may be required and implemented yet this may conflict with emerging evidence, resources available, and priorities of those impacted by, and responding to, shocks. Whilst this may be considered essential, this can conflict with other priorities and the design of sound, long-term processes that enable resilience to be implemented and sustained. Climate services, for example, provide timely information, translated and delivered in order to best inform decision-making in the context of climate mitigation and adaptation (National Research Council 2001) thus providing a process through which both reactive and proactive approaches to nexus shocks can be undertaken. Whilst this ensures a closer alignment with the needs of end-users and those who require climate information to inform their responses, climate services are limited by a narrow range of products and ways in which information is provided, made available by scientists and a lack of consideration of societal responses to climate change (Brasseur and Gallardo 2016).

Challenges that affect effective decision-making in response to nexus shocks include access to appropriate data or fragmented information (Soares et al. 2017), limited applicability for on-the-ground decision-making particularly for shock events and resource constraints (Bhave et al. 2016), all of which affect the ability for long-term planning and resilience to nexus shocks. In addition, data collection, storage and analysis, capacity building and end-user connectivity need to be central to these approaches (Vaughan et al. 2016). Conversely, adopting processes of decision-making that are robust and effective present a number of opportunities, such as increased urbanization and transformative adaptation, improved forest and conservation management and benefits to disaster risk management approaches (Bhave et al. 2016). Whilst many of these challenges are related to information and data availability, access and applicability, the provision of more information is not always or necessarily the answer. This is particularly true when considering how adaptive capacity building ensures related attributes, such as information, expertise, leadership and organizational culture (among others) are also considered. Experience and context are important components to consider in these responses to nexus shocks, this is especially true given the complexities, unpredictable, uncertain nature of the interactions (which at times can be conflicting) of nexus resources (i.e. energy, food, water, environment), sectors that depend on them and stakeholders that interact within them (e.g. businesses, communities, policymakers, citizens). It is, therefore, necessary to adopt a reflective approach whereby there exists a deep understanding and appreciation for the factors that make shocks worst (i.e. exacerbate) or reduce their impact (i.e. mitigate), or space for effective collaborative partnership that brings in relevant knowledge from relevant parties.

The factors exacerbating and mitigating shocks discussed in this chapter are categorized according to the five themes explored in the Nexus Shocks workshops, described in Chapter 1: Predicting shocks and hazards, Transmission and mitigation of nexus risks though infrastructure, Local economy responses to shocks, Finance and insurance for resilience, and Governance and governments. The findings of these workshops are presented and discussed here and these do at times present similarities suggesting that what may be perceived as exacerbating a shock from one perspective (i.e. data and evidence provision) may be a way of mitigating the impacts of a shock from another (i.e. governance and governments). Participants represented universities, UK local and national government departments, city-based climate initiatives, Non-Governmental Organizations, businesses, finance organizations, consultancies and climate media/communications agencies. Consequently, the sample is likely to reflect the views of individuals and organizations who have the direct or indirect knowledge and experience of working on decision-making processes in response to climate-and weather-related shocks. Each workshop was conducted under the Chatham House rule whereby 'participants are free to use the information received, but neither the identity nor the affiliation of the speaker(s), nor that of any other participant, may be revealed'.

THE PREDICTION OF CLIMATE AND WEATHER SHOCKS

There is growing literature in science studies which stipulates that scientific evidence alone cannot drive policy (Sarowitz 2004; Jasanoff 2012). Rather than producing constructive solutions to environmental challenges, Lövbrand and Öberg argue that scientists charge themselves with the 'impossible task of reducing uncertainties and providing irrefutable truths in complex environmental controversies' (2005: 195). A study published in 2014 examined the translation of scientific evidence into policymaking on climate change at the local level through the UK National Indicators and concluded that the science-policy process 'excluded local knowledge about both the contexts for [emissions] and the ability of local authorities to exercise control over the sources of such emissions' (Pearce 2014: 198). The model of 'truth speaks to power' where scientific evidence underpins decision-making, remains linear and as such is outdated considering the modern non-linear, human-driven challenges society faces today such as shocks from climate change to the food, energy, water, environment nexus. But rather than evolving, a lack of a reflexive approach and philosophy which informs discussion and responses, has led to a self-sustaining and reinforcing cycle, where 'disciplinary orientations embody different normative assumptions and divergent notions of good and relevant science' (Lövbrand and Öberg 2005: 196).

To explore this further, a first workshop was run in collaboration with the UK Met Office which has responsibility for predicting extreme weather events and their impacts and whose role is to forecast weather and model climate change for the UK and worldwide. It provides world leading weather services for the public, businesses and government, making it a leading knowledge provider to decision-makers. Where a nexus shock is at risk of occurring, the Met Office has a direct line to inform stakeholders ranging from those at the household level to the Government, the Cabinet Office, recovery units and the media, to advise them on their responses. Being involved with the Nexus Shocks project provided the Met Office with the opportunity to share its lessons and experiences with other knowledge providers, as well as practitioners. It gave them the chance to explain how they go about addressing the complexity and uncertainty associated with shocks to the nexus, particularly in relation to their communications strategy.

Constructing responses to nexus shocks requires sound evidence and advice from a range of experts. The *timing* of the provision of this evidence is crucial to the process but is both an advantage and a limitation. Evidence feeds into pre-warning systems which enables timely evacuation in shock situations particularly in vulnerable areas, but conversely the inability to communicate the nature and timing of these shocks, or to communicate the risks associated with them present huge limitations when this is needed on the ground. A lack of understanding of the *culture* of scientific processes (and indeed the policy making process), the production of knowledge and how science is used to inform decision-making processes (and the social factors that influence this) can further limit how evidence effectively informs responses to nexus shocks. However, this uncertainty can also provide a degree of flexibility in producing alternative scenarios of impacts and responses that can each or simultaneously underpin decision-making processes. The *evidence provided* and the way it is communicated can lead to different interpretations and it having different meanings for different stakeholders who use it in the aftermath of a shock, such as local communities. However, having *adequate communication processes* in place enables a better understanding of the needs of those who use evidence to inform their decision processes and ultimately enables tailored messages to be disseminated clearly, concisely, and in an informative way to those audiences to ensure appropriate evidence can inform appropriate response measures. For example, the UK Met Office adopts a traffic light system when dealing with the media and public to highlight different levels of risk and in order to warn both the public and emergency services of extreme weather which may have potential disruptive or life-threatening effects. This red, amber, yellow system (red: highest impact; amber: medium impact; green: minimal impact) is tailored to a particular event based on the likelihood of it occurring and the potential impacts it can lead to and are shared through various media: TV, radio, online, social media, mobile phones, RSS feeds and email alerts. By focusing on the impact rather than on the science, the information, therefore, becomes relatable and degree of urgency become obvious, although this not be applicable across all stakeholder or ethnic groups. *Credibility* of the organization is key in delivering the messages to ensure consistent and trusted information exchange; in the UK, the Met Office is a highly respected and credible source of advice and science. Similarly, failure to adequately or accurately predict a shock (or to predict a shock which does not materialize) can affect this credibility and lead to a 'crying wolf' phenomenon where no response can occur after a shock if this has been mis-predicted in the past. An example of a false warning occurred in January 2018 in the State of Hawaii, USA, when the Hawaii Emergency Manager Agency issued a false ballistic missile alert through its Emergency Alert systems. It took over 30 minutes to rectify the alert and inform citizens that this was a false alarm (FCC 2018) and whilst it led to widespread but minimal damage it may have caused future risks of citizen apathy should a similar shock event occur.

The Relationship Between Shocks, Their Impacts and Infrastructure

Any country's progress in terms of health, wealth and security relies upon critical infrastructure, such as transport systems, renewable and non-renewable energy generation plants, industry, water supply networks and education and health infrastructures each of which exhibits different degrees of vulnerability (Forzieri et al. 2018) to shocks. Transport infrastructure is particularly vulnerable to extreme weather events, temperature extremes (icy weather, drought) and increased precipitation leading to growing pressure on governments to ensure the functioning of transport networks and their resilience in the long term (Kiel et al. 2016). Transport infrastructure may appear outdated and unable to cope with the impacts of nexus shocks both in terms of its ability to facilitate pre- and post-shock emergency responses (such as evacuation processes mobilizing a high number of people in a short period of time). Its low resilience to the growth in anticipated shocks may also lower the ability of systems and processes, such as emergency services, which depend on transport to recover from shocks (Beheshtian et al. 2018).

A workshop was co-designed with Atkins, an international engineering, design and management consultancy, considered a key stakeholder in infrastructure development from the local to the global level. Atkins' work on resilience means it is inherently interested in nexus shocks which fed into the design of the workshop. Decision-making associated with infrastructure occurs throughout its lifecycle and must have a response to these types of shocks embedded within them, minimizing the risk to infrastructure itself and the services it provides. Different types of shocks cause different impacts and implications for infrastructure. For example, systemic risks can occur when a shock travels from one asset or infrastructure system to others having a knock-on effect through supply chains. Slow shocks are when no individual event is severe but the cumulative impact of 'mini shocks' leads to a tipping-point whereby the overall impact is significant. The geographic location and spread of a shock can impact a whole country, for example, meaning any response is thinly spread over a wider area and hinders the effectiveness of recovery. Continental shocks, on the other hand, occur in an ever-connected and globalized world where a shock on one area of the nexus can cascade across the nexus and beyond national boundaries. Finally, the overarching vulnerability and connectivity of shocks is important to consider, where a shock in one part of the world can be connected to a shock in another, which can lead to additional vulnerability in supply chains with severe implications across multiple nexus stakeholders, sectors and resources.

Resilience of infrastructure such as buildings and transport networks, is supported by long-term and strategic *investment decisions*, but these are heavily affected by any lack of clarity that may exist on the purpose and structure of an asset in question and how this asset may affect the flow of resources at the time of a shock. Current and future owners of assets can, therefore, affect responses to nexus shocks if there is a lack of access to revenue to support the development and maintenance of these assets. However, current and future asset owners can also affect decision-making by fully understanding the resilience of an asset and making a clear distinction between what is meant by the value of such an asset (e.g. intrinsic, to society) and its cost. The nature of an asset can affect the resilience, vulnerability and ability of infrastructure as a whole to withstand shocks to the nexus, and when such a shock occurs if the asset itself is too weak, this may lead to failure of responses, longer time lags for the system to recover and an exacerbation of inherent vulnerabilities. This has repercussions for the hierarchy of measures that may be used post-shock to ensure infrastructure survival and resilience. This is especially relevant when considering the tradeoffs between quickly restoring infrastructure services in the aftermath of a shock against strategic, longer-term approaches that consider alternative approaches that may be available yet extend a period by which such an asset is unavailable (MacAskill and Guthrie 2015). These independencies of shocks and affected infrastructure can also create opportunities to identify areas of highest resilience and build this into broader mechanisms for enhancing resilience and reducing vulnerabilities across a system. The rapid pace at which society is depending on *Intelligent Communications Technologies*

(ICT) is an example of this, creating new and 'unknown unknowns' on which limited evidence exists of how shocks affecting ICT infrastructure could have societal implications, even with the benefits of technology enhancement enabling increased utilization of assets.

The *social dimension* of infrastructure holds an important position in informing responses and resilience to shocks, particularly as societal progress is reliant upon the resilience of infrastructure which facilitates interactions across energy, food, water, environment sectors and stakeholders. For example, without adequate and resilient transport infrastructure in order to transport water to cities and support agricultural practices or energy infrastructure to supply power to communities and stakeholders, responses to climate shocks to the nexus will not fully adhere to the needs and demands of society, which themselves alter following a shock. The unpredictability of shocks means that societal responses can be sporadic or unpredictable leading to a possible increased likelihood that these responses will be reactive rather than proactive. This can lead to further unexpected and unanticipated impacts exacerbating the original shock that occurred. For example, forecasts for incidents of heavy snow which may not have been predicted with sufficient time, may result in transport infrastructure not being adequately cleared in time for schools to open, hence these remain shut and children are forced to stay home with their parents who as a result are unable to go to work, having further implications for economic growth. The March 2018 heavy snow in the UK for example is estimated to have cost the economy up to £1 billion per day (BBC News 2018). However, proactive response strategies have been put in place specifically to address issues of societal implications of a shock event characterized by being a short-term threat; these responses can be both short and long term in nature depending on the initial shock, impacts and resilience of the system.

Coping mechanisms put in place, such as the London Adaptation Pathways (see Chapter 1), are based on collaboration to ensure all potential scenarios are considered and a range of options for responses are provided. When it comes to infrastructure, and updating existing assets which may be more vulnerable to shocks due to their age or use, these Adaptation Pathways enable a forward-looking mechanism to build on lessons learnt and ensure that future designs are useful, suitable for future contexts and turn issues that exacerbate impacts of climate shocks to the nexus into mitigating ones. This is even more relevant when considering implications of *competing demands* when continuous efforts to optimize a system and increase the efficiency of the current infrastructure can increase the likelihood of shocks cascading through the food, energy, water, environment nexus and further increase infrastructure vulnerability to the shock. In this case, inappropriate or mis-*communication* around the vulnerability and potential impact to all stakeholders involved can further exacerbate the original impacts of the shock. This is further complicated if the climate shock impacts and disables energy infrastructure upon which communication channels depend leading to alternative, or more classic (e.g. radio) means of communication becoming necessary. Learning from these incidents enables the improvement of the communication of practical information on how to respond in addition to whom to communicate key guidance to in order to pre-empt possible unanticipated consequences.

Experiencing Shocks at the Local Level

A robust evidence base needs to be salient, credible and legitimate (Cash et al. 2002), available and easily accessible to actors across sectors and governance boundaries to enable judgement-based decisions (Twigger-Ross et al. 2015) reflecting a sound understanding of the knowledge needs of different stakeholders (Bruine de Bruine and Bostrom 2013). The lack of provision of such information can be further exacerbated by 'conflicting timescales between research and policy combined with the social dimensions of decision-making and the need for researchers to achieve consensus before they can contribute to decision-making' (Howarth and Monasterolo 2016: 58). Shocks inherently require quick and effective decision-making, and therefore, timescales between actors need to be planned and aligned in advance. At the local level, a key observable response to the risk of food price spikes in the UK in the past 10 years has been an increase in the number of community food growing initiatives, the main aim generally being to ensure a sustainable, affordable, healthy, high quality and accessible food supply rather than focusing on the reliance and availability of exports and imports, food security, or reducing energy or water usage (White and Stirling 2012). Collaboration and communication between the public sector and civil society play an important role in the effectiveness of both top-down and bottom-up decision-making in response to nexus shocks through the provision of information, voluntary action and joint implementation of local projects (Nasiritousi et al. 2016). Decisions at the national level may fail to consider implications for the local level or in the long term, whole-sector implications of nexus shocks and grassroots participation from local communities is needed to design sustainable and resilient responses to these shocks (Howarth 2016). A robust process is therefore needed to '[synthesize] evidence to inform local decision-making on climate change whilst simultaneously drawing on more localized expertise' (Howarth and Painter 2016: 10).

The third workshop featured in this chapter was co-designed and run with Cambridge Cleanteach and LDA Design. Cambridge Cleantech is a stakeholder in the Cambridge local economy, supporting its members by exploring global opportunities for technology innovation and linking these with companies in the sector. Shocks to the nexus present opportunities for emerging technologies, especially in how they can provide some solutions to mitigate against shocks in terms of preparedness and how they can support recovery post-shock. LDA Design, an independent landscape, urban design, environment and planning consultancy, demonstrated that shocks to the nexus have the potential to affect us all, unless we take action to build resilience in the built environment, infrastructure and the landscape. This builds on work they coordinated to explore the major economic, social and environmental drivers of change in the world and the challenges and opportunities they present, setting out practical solutions which can deliver results and add value for clients.

Societal responses to nexus shocks, as discussed above, can exacerbate impacts of shocks particularly when behavioural responses can act as catalysts for these shocks. In these instances, vulnerability and sensitivity to a shock can be made worst when members of the public inadvertently put themselves in harm's way in the immediate aftermath of a shock, for example in order to reach loved ones, or as a result of having not heard local warning systems and associated guidance. This means additional emergency efforts may need to be deployed specifically to aid them in addition to broader efforts to support local communities. A way to mitigate this is by being selective and constructing reflective local *communication* approaches to ensure appropriate, timely and targeted information is provided to those who need it within required timeframes. This should contain locally relevant information which can enable individuals and communities to ascertain how they can contribute to a resilient societal response to climate shocks. Similarly, local businesses need to be equipped with the right knowledge, skills, capacity and resources to boost the resilience of their business whilst acting as trusted voices and relaying key guidance on responses to climate shocks to citizens. *Collaboration* is equally an enabler to sustainable responses to climate shocks. Working in silos can exacerbate climate shocks, as local businesses and local authorities may not be able to see how their individual roles fit into the broader picture of responses, or understand how they are impacted or their counterparts may be impacted, or how their actions may impact others or exacerbate existing impacts experienced by others. For example, the health sector plays an important role in identifying, assessing and acting on climate risks and how these may impact the health of vulnerable individuals, however, with little or no collaboration with other stakeholders including policymakers, health professionals cannot alone alleviate these impacts or design policies and approaches to reduce them (Machalaba et al. 2015). Putting in place sharing mechanisms will enable companies to work together at the local level to maximize resilience, foster innovation and creativity and forge lasting, trusted, collaborative and transparent partnerships. Local communities are an often untapped, yet important, resource upon which responses need to based, failure to include them in plans for responses can further exacerbate impacts whilst they provide a unique resource and window into the resilience needs and abilities of their members. Collaborative processes thereby enable a better flow of insights, evidence and research around all possible scenarios of impacts, vulnerabilities, sensitivities and potentials for resilience thereby increasing understanding of possible consequences, unanticipated impacts and alignment of resilience design with this.

Whilst collaboration can overcome the threat of multiplying characteristics of climate shocks to the nexus it can bring to light issues arising from the *organizational culture* that exists within a system impacted by a shock. Due to the variety of sectors and stakeholders affected by a climate shock and further enhanced vulnerabilities and affected actors working across the food, energy, water, environment nexus, a web of organizational mechanisms can slow responses or impede longer-term sustainable practices. A focus on immediate profit and short-term benefits, often driven by cost-benefits instruments used by policymakers and businesses, do not always take into account future costs of inaction. For example, immediate responses to flood events could entail the deployment of sandbags and emergency responses, yet longer-term measures such as reviewing bridge or road infrastructure damaged by flooding, may be overlooked. Such a reactive approach is systemic within individuals and local businesses whose planning period may not extend beyond a few months and may not factor in longer-term projections of climate shocks and their local impacts. The ways in which individuals and businesses plan differs considerably at the local level where a lack of future-thinking is apparent and an over-reliance on reactive responses exists, particularly with some of the decadal timescales within which the intensity and frequency of nexus shocks is likely to occur. Whilst these can make the impacts of nexus shocks worst, they also provide an opportunity to better understand the needs of stakeholders and end-users who operate locally. This provides space for a process through which messages to inform on predicted shocks, possible impacts and advice on responses, can be properly tailored to a particular shock and audience and simplifying this particularly around specific decision-making points.

THE ROLE OF FINANCE AND INSURANCE

Climate change is likely to lead to more frequent and intense weather extremes such as extended periods of dry weather and precipitation leading to more pronounced heatwaves and floods. The finance and insurance industries have incurred significant financial losses as a result of extreme weather and climate events as the industry is both impacted in terms of invested assets experiencing the impacts of climate change and insurance costs associated with those impacts. The insurance industry is impacted by climate shocks through physical risks (such as flooding and storms), transition risks (resulting from transitioning to low carbon economies) and liability risks (i.e. when stakeholders may make claims to cover damages they have incurred) (Prudential Regulation Authority 2015). The last decade (2007–2017) in particular has had more frequent and disruptive climate and weather extremes representing 94% of insurance claims globally in 2015 (Munich Re 2016) and in 2017, insurance claims amounted to $330 billion, only overtaken by the claims for 2011 which exceeded $350 billion (Munich Re 2018). They are seen as obvious investors in resilience, and the insurance industry has significant opportunities to do so, although this may not be their primary priority (CISL 2016) and catastrophe modelling is increasingly used across the finance and insurance industries in order to identify and underwrite certain risks and assess asset exposure (Lloyds 2014).

A workshop was run in collaboration with Lloyd's of London and Willis. Lloyd's of London is the world's specialist insurance and reinsurance market, specializing in new, unusual and complex risks, and they recognize that complex impacts arising from extreme events have the potential to significantly impact the insurance industry. Lloyd's has previously looked into nexus shocks, specifically considering a shock to the global food system in a report released in 2016. This work used scenario analysis to better understand impacts of what might happen—as opposed to what will happen—in the case of an acute shock to world food supply. Willis is a multinational risk advisor, insurance brokerage and reinsurance brokerage company. Shocks—or the risk of shocks—to the nexus exacerbate existing vulnerabilities caused by financial markets, regulatory change, political risk, resource stress and technological and environmental change. Many of these current and future risks are unrecognized by owners and operators and continue to be uninsured; this causes a resilience deficit, for insurance is integral to smoothing the financial impacts of large losses caused by shocks.

Whilst the processes initiated following a nexus shock are known within the finance and insurance industries, and incorporated into decision-making tools on investment and insurance, there is a widespread *lack of understanding* of the interdependencies within shocks themselves. The food, energy, water, environment nexus is fraught with intricacies and complexities, given how stakeholders may depend upon or work within more than one resource, how decisions affecting one sector may have implications for another, and the aims to maintain the resilience of one sector or resource may conflict with those for another. Adding to this the unpredictability of occurrence and impact of shocks to the nexus and uncertainty about scale, degree of impact and responses, further complicates the interdependent nature of the nexus and the shocks that affect it. The consideration of the *scale* of shocks and their impacts highlights the lack of adequate targeting of responses. When there is confusion or lack of understanding of which stakeholders and sectors are affected at the local, national or global levels, and whose responsibility it may be to lead responses, it is a challenging feat to establish appropriate means for designing, implementing and evaluating these responses. Analyzing these shocks from the perspective of a current or future asset owner can provide greater insight into how decision-making processes are affected by present and future shocks across a number of scales and nexus resources. Adopting a lens by which value and cost are clearly distinguished can enable the complexities and uncertainties that emerge from examining shocks at particular scales to be dampened and move away from being an exacerbator of shocks to an obstacle that can be overcome.

The nature of financial decision-making is designed and conducted on different *timescales* to those through which a shock manifests itself, for example, insurance programmes are written on an annual basis with no real long-term process in place. When considering investments, these can be made on timescales ranging from a second to a year and up to five to seven years for private equity or infrastructure. This short-termism presents significant limitations when aligning decision-making processes with predictions of frequency and intensity of shocks that are likely to occur and affect societal systems and the assets upon which they depend. This entails moving away from short-term thinking with stronger consideration for the life of assets with a view to embedding deeper long-term and strategic thinking where the operational lifetime of an asset enables short-term considerations to be embedded within longer-term planning and investment. Financial decision-making often focuses on the profit rather than the cost of investing in innovation, and this is further supported by public sector subsidies into insurance which are aimed at keeping costs to a manageable level. However, this can lead to investment into projects that are already vulnerable and lack properties that will ensure they are sustainable in the longer term. An example of this is the joint industry and government sponsored scheme *Flood Re Transition Plan* (Flood Re 2016) which focuses on supporting the transition to a 'market with risk-reflective pricing' in order to ensure affordable insurance cover to households most vulnerable to and at higher risk of flooding. Whilst this has reduced the cost of flooding of those at very high risk in high flood-prone areas it does not account for homes in flood-prone areas that are at a lower risk of flooding. Similarly, social responses to insurance available in case homes and assets are affected by flooding may lead to a *moral hazard* where this safeguard may reduce salience and increase apathy towards the issue, as people are less aware of the risk of their exposure, affecting longer-term thinking into ensuring sustainable approaches to improving the resilience of a given asset. Indeed, Bubeck and colleagues suggest that a strong sense of security can emerge from purchasing flood insurance resulting in limited mitigation behaviour (Bubeck et al. 2012), and a team of academics in the United States found that households experiencing recent flood damage positively impact decisions to purchase flood insurance, however, this signal disappears after three years (Atreya et al. 2014). Adopting a proactive approach by assessing and identifying which assets are most at risk and engaging directly with insurers can enable a more constructive and sustainable mechanism through which the financial impacts resulting from nexus shocks to business disruption, for example, can be better explored and accounted for in decision-making processes.

Finance available cannot itself offer a price discovery mechanism, which is based on the interaction of buyers and sellers in the market, if the sector as a whole and the stakeholders within it, are not fully engaged in identifying the risks at stake and how these can be modelled appropriately. These interactive processes being in place ensure up-to-date price signals are shared with decision-makers to further inform responses in the short and long term. *Lessons* on this should be better captured, for example, from interactions in developing countries where insurance companies are more prone to working in partnership with stakeholders across the food, energy, water, and environment nexus. Progress is being made within this context particularly where new models of insurance in nexus shock situations are being developed, such as parametric insurance which makes a payment in the occurrence of a specific shock characterized by low frequency and high impact, such as hurricanes or earthquakes, as opposed to waiting for economic value of a loss to be assessed in the longer term.

GOVERNANCE AND GOVERNMENTS IN SHOCK EVENTS

The information-deficit approach assumes that scientific evidence and advice informs policy-making and decisions, and where this assumes 'truth speaks to power' remains the main framework through which scientists and policymakers interact (Jasanoff and Wynne 1998). This approach assumes that providing more information is sufficient to ensure decisions align with the range of complexities and uncertainties that the original information may contain. Whilst it is a process that has served well historically, it is limited firstly by its inability to evolve, reflect and align with more modern and complex societal challenges, such as climate shocks and their impacts to the food, energy, water, environment nexus and the systems across society that are affected by this. Second, the expectations of those making decisions and designing policies combined with an over-reliance on evidence-based policy-making cannot be fulfilled as these shocks are fraught with large uncertainties. In addition, decision-making processes are highly dependent on context where 'more than one outcome is consistent with expectations' (Dessai et al. 2009: 67). As a result, a number of uncertainties that exist and emerge from decision-making originate from the vast range of options and evidence available combined with the multitude and complex nature of the processes in place to decide between them (Brunsson 2007). And as such, any limitations that may become apparent in the production of evidence on climate risks, impacts to the nexus and related responses may not necessarily reflect or be interpreted as limits to subsequent decision-making, particularly when large uncertainties in this evidence exist. Decision-making and response to shock events are very much dependent on the context within which they occur and decision-makers operate. Decision-makers are also not always able to take on board the volume and complex nature of the evidence they are presented with and the randomness and unpredictable nature of interactions between them, solutions and alternatives further complicates the process.

A workshop was designed in collaboration with the Energy, Environment and Resources Department (EER) at Chatham House which seeks to advance the international debate on energy, environment and resources policy and enable decision-makers and governments, NGOs, business and the media, to make well-informed decisions and mitigating potential future climate and resource-related insecurities. Chatham House's EER department interest in shocks to the nexus came from research undertaken on critical chokepoints in global food trade to develop sustainable, risk-based strategies for global food security (Bailey and Wellesley 2017). This required an understanding of the vulnerability of global trade to disruptions of all kinds: the food system must be resilient not only to shocks within the food supply chain—such as harvest failures or import and export restrictions—but also to those resulting from broader trade, political, and environmental dynamics.

Social responses to nexus shocks can further exacerbate a shock where the perception of risk can lead to significant unrest. This can be mitigated by liaising closely with communities and stakeholders working with the public to ensure appropriate messages are communicated, expressing the current and projected levels of risk alongside advice on suggested behavioural action to take. Close links between the public, local organizations, communities, networks and trusted actors can facilitate a process of effective and rapid communication, ensuring dissemination of most up-to-date information via appropriate channels during a shock. Consistent mechanisms for dissemination must also be maintained in the aftermath of a shock, particularly if this has impacts for those working across and affected by resources within the nexus, to ensure long-term thinking and sustainability of a system are maintained and that, should future shocks occur, such a process is already trusted and familiar to its users. *Governance processes*, or 'processes of interaction and decision-making among actors involved in a collective problem that lead to the creation, reinforcement, or reproduction of social norms and institutions' (Hufty 2011), rely on mechanisms of power which are distributed across the local, regional, national and international scales. This web of decision-making levels can demonstrate a lack of clarity as to who owns a problem emerging from a nexus shock (and hence who is most affected by it) and who is responsible for driving solutions to alleviate the problem (which may differ from the problem owner). Badly coordinated responses to extreme events or cumulative shocks can thereby lead to a lockdown of action, particularly when a division of portfolios across departments within one government can make it difficult to adopt a coordinated approach. In spite of this, however, legislation can survive changes in government—meaning that robust decision-making processes in place to respond to a nexus shock can be altered and updated with time, irrespective of a government party in power.

Strategic responses are required across governments but these can be sparse where a different response may emerge depending on which government department is engaged. For example, the Department for Food, Environment and Rural Affairs, which prioritizes the safeguarding of the UK's natural environment and food and farming industry may have different priorities to focus on in the aftermath of a shock than the Department for Business, Energy and Industrial Strategy. This suggests that, in the context of a nexus shock, the roles of different government departments may not be as clear as expected, their priorities may in fact conflict and hence how they work together lacks a distinct strategy. In the UK, the broader function of government is to ensure national security is preserved, so, in the context of a nexus shock, such a government is obligated to respond. The UK government operates in such situations through its Cabinet Office Briefing Rooms (COBR) to respond to national crises in a timely manner, where there exist emergency funding mechanisms, risk management regimes and contingency plans to put in place. The allocation of responsibility within a limited *timeframe* available to respond to a shock can raise challenges as due to the very nature of the shock, which occurs irrespective of professional or political priorities, the most appropriate resource or leadership may not be in place at the time of the shock.

Timescales by which governments construct responses to shocks can exacerbate initial impacts. Different lag times exist for different types of responses and stakeholders involved in decision-making processes, and if a government's reaction occurs within a short timeframe (e.g. aligning with a five-year political term for example) this can increase the initial impact of the shock. For example, wheat exports from Russia, Ukraine and Kazakhstan represent about 20% of grain traded in the World's market and this is predicted to grow significantly (OECD-FAO 2012). However in the periods 2007–2008 and 2011–2012, grain harvests in these countries, affected by severe droughts, led to restrictive export bans. These export restrictions have been found to have exacerbates the original impacts of the drought that occurred leading to increased grain prices with projections of price increases up to 23% for wheat globally as a result of severe droughts. Accounting for these different impacts which may have longer timeframes than original ones within which a nexus shock occurred, can help alleviate broader, cascading impacts of nexus shocks.

LEARNING FROM EXACERBATING AND MITIGATING FACTORS

Some of the key factors that exacerbate the damaging impacts of nexus shocks also act to mitigate them: timing, collaboration, evidence, culture, responsibility and responses, and technology. Failure to adequately predict when shocks may occur or impacts they will have can negatively affect credibility when there are limited time and resource at a time when a shock occurs or resources are used and funding spent when no shock manifests itself. This can lead to overly precautious approaches or an over-reliance on external safeguards, such as insurance and others to take the lead. Timing of the provision of evidence to inform decision-making (or the lack of accuracy in predicting a shock and its characteristics) in a way that is relevant and of use to stakeholders who are required to formulate responses, can result in a misunderstanding of the risks involved. Many of these responses involve collaboration across a number of stakeholders (which we explore in Chapter 4) but the involvement of multiple individuals and organizations can lead to competing demands and potential increases in the likelihood of shocks cascading through other sectors within the nexus ultimately increasing the vulnerability of actors, resources, processes or infrastructure within the nexus. Technology is a useful tool in predicting and formulating responses to nexus shocks, however, an overdependence on these tools can create conflicting and possibly unknown vulnerabilities in the system. Attributing responsibility, owning leadership and retaining credibility are all at risk when formulating responses to nexus shock events, particularly when a shock is not by its nature static and can exhibit impacts that affect more than one sector, stakeholder or policy issue.

Factors that help mitigate detrimental impacts of nexus shocks range from needing to clarify costs, leadership roles, communication and collaboration, and clarity on timescales. Nexus shocks are often associated with negative impacts and associated costs, and they are likely to result in multiple impacts leading to a multitude of financial and non-financial (i.e. loss of life) costs. These are important to consider as both can help frame decision-making and lead to unanticipated benefits creating opportunities to build resilience and effective sustainable decision-making to nexus shocks. The unpredictable and uncertain nature of nexus shocks can lead to strong internal leadership within organizations and increased engagement with other sectors and stakeholders to better respond to these shocks. Increased communication of evidence and impacts to specific audiences, combined with efforts to strengthen capacity building, may raise resilience and awareness, strengthen collaboration across stakeholders and build partnerships across sectors. The broad consideration and reflection on the impacts of shocks can lead to shifts in timescales with a more sustainable balance between short-termism and long-term thinking.

References

Atreya, A., Ferreira, S., & Michel-Kerjan, E. M. (2014). What drives households to buy flood insurance? New evidence from Georgia. *Ecological Economics, 117,* 153–161.

Bailey, R., & Wellesley, L. (2017). *Chokepoints and vulnerabilities in global food trade* (Chatham House Report).

BBC News. (2018, March 5). *Reality check: Did snow cost the UK economy £1bn a day?* Available online at http://www.bbc.co.uk/news/business-43287975.

Beheshtian, A., Donaghy, K. P., Gao, H. O., Safaie, S., & Geddes, R. (2018). Impacts and implications of climatic extremes for resilience planning of transportation energy: A case study of New York City. *Journal of Cleaner Production, 174,* 1299–1313.

Bhave, A. G., Conway, D., Dessai, S., & Stainforth, D. A. (2016). Barriers and opportunities for robust decision-making approaches to support climate change adaptation in the developing world. *Climate Risk Management.* http://dx.doi.org/10.1016/j.crm.2016.09.004.

Brasseur, G. P., & Gallardo, L. (2016). Climate services: Lessons learned and future prospects. *Earth's Future, 4,* 79–89. https://doi.org/10.1002/2015ef000338.

Bruine de Bruine, W., & Bostrom, A. (2013). Assessing what to address in science communication. *Proceedings of the National Academy of Sciences USA, 110*(3), 14062–14068.

Brunsson, N. (2007). *Consequences of decision-making.* Oxford, UK: Oxford University Press.

Bubeck, P., Botzen, W. J. W., & Aerts, J. C. J. H. (2012). A review of risk perceptions and other factors that influence flood mitigation behaviour. *Risk Analysis, 32,* 1481–1495.

Cash, D., Clark, W., Alcock, F., Dickson, N., Eckley, N., & Jäger, J. (2002). *Salience, credibility, legitimacy and boundaries: Linking research, assessment and decision-making* (Harvard University Faculty Research Working Papers Series). Cambridge, MA: Harvard University.

De la Fuente, A. (2007). *Climate shocks and their impact on assets.* Human Development Report 2007/2008 Fighting climate change: Human solidarity in a divided world (Human Development Report Office Occasional Paper 2007/23).

Dessai, S., Hulme, M., Lempert, R., & Pielke R., Jr. (2009). Climate prediction: A limit to adaptation? In N. W. Adger, I. Lorenzoni, & K. O'Brien (Eds.), *Adapting to climate change: Thresholds, values, governance* (p. 515). Cambridge: Cambridge University Press.

Federal Communications Commission. (2018, January 30). *Preliminary report: Hawaii Emergency Management Agency's January 13, 2018 false ballistic missile alert.* Public Safety and Homeland Security Bureau. Available at https://www.fcc.gov/document/presentation-preliminary-report-hawaii-false-emergency-alert.

Flood Re. (2016, February). *The first Flood Re transition plan: Transition to an affordable market for household flood insurance.* Available at http://www.floodre.co.uk/wp-content/uploads/Flood-Re-Transition-Plan-Feb-2016-FINAL.pdf.

Forzieria, G., Bianchi, A., e Silva, F. B., Herrera, M. A. M., Leblois, A., Lavalle, C., et al. (2018). Escalating impacts of climate extremes on critical infrastructures in Europe. *Global Environmental Change, 48,* 97–107.

Howarth, C. (2016). *What we've learnt so far: Findings from the Nexus shocks network.* Global Sustainability Institute.

Howarth, C., & Monasterolo, I. (2016). Understanding barriers to decision making in the UK energy-foodwater nexus: The added value of interdisciplinary approaches. *Environmental Science & Policy, 61,* 53–60.

Howarth, C., & Painter, J. (2016). The IPCC and local decision making on climate change: A robust sciencepolicy interface? *Palgrave Communications, 2,* 16058. http://dx.doi.org/10.1057/palcomms.2016.58.

Hufty, M. (2011). Investigating policy processes: The Governance Analytical Framework (GAF). In U. Wiesmann, H. Hurni, et al. (Eds.), *Research for sustainable development: Foundations, experiences, and perspectives* (pp. 403–424). Bern: Geographica Bernensia.

Jasanoff, S. (2012). *Science and public reason.* New York: Routledge.

Jasanoff, S., & Wynne, B. (1998). Chapter 1: Science and decision making. In S. Rayner & E. L. Malone (Eds.), *Human choice and climate change* (pp. 1–87). Columbus, OH: Batelle Press.

Kiel, J., Petiet, P., Nieuwenhuis, A., Peters, T., & van Ruiten, K. (2016). A decision support system for the resilience of critical transport infrastructure to extreme weather events. *Transportation Research Procedia, 14,* 68–77.

Lloyds. (2014). *Catastrophe modelling and climate change* (41 p.). London, UK: Lloyds.

Lövbrand, E., & Öberg, G. (2005). Comment on "How science makes environmental controversies worse" by Daniel Sarewitz, *Environmental Science and Policy, 7,* 385–403 and "When scientists politicise science: Making sense of the controversy over the skeptical environmentalist" by Roger A. Pielke Jr., *Environmental Science and Policy, 7,* 405–417. *Environmental Science & Policy, 8,* 195–197.

MacAskill, K., & Guthrie, P. (2015). A hierarchy of measures for infrastructure resilience—Learning from post-disaster reconstruction in Christchurch, New Zealand. *Civil Engineering and Environmental Systems, 32*(1–2), 130–142.

Machalaba, C., Romanelli, C., Stoett, P., Baum, S. E., Bouley, T. A., Daszak, P., et al. (2015). Climate change and health: Transcending silos to find solutions. *Annals of Global Health, 81*(3), 445–458.

Munich Re. (2016). *Natural catastrophes 2015: Annual figures.* Munich Re NatCat Service. Available at http://www.munichre.com/site/corporate/get/params_E-1254966961_Dattachment/1130647/Munich-Re-Overview-Natural-catastrophes-2015.pdf.

Munich Re. (2018, January 8). *Natural catastrophe review; series of Hurricanes makes 2017 year of highest insured losses ever.* Munich Re Press Release. Available at https://www.munichre.com/en/media-relations/publications/press-releases/2018/2018-01-04-press-release/index.html.

National Research Council. (2001). *National Research Council, 2001. A climate services vision: First steps toward the future.* Washington, DC: National Academies Press.

Nasiritousi, N., Hkerpe, M., & Linner, B.-O. (2016). The roles of non-state actors in climate change governance: Understanding agency through governance profiles. *International Environmental Agreements: Politics, Law and Economics, 16*(1), 109–126.

OECD-FAO. (2012). *OECD-FAO agricultural outlook 2012–2021.* Organisation for Economic Cooperation and Development (OECD), Paris, and Food and

Agricultural Organisation of the United Nations, Rome, OECD Publishing and FAO.

Pearce, W. (2014). Scientific data and its limits: Rethinking the use of evidence in local climate change policy. *Evidence and Policy, 10*(2), 187–203.

Prudential Regulation Authority. (2015). *The impact of climate change on the UK insurance sector* (A Climate Change Adaptation Report by the Prudential Regulation Authority). London, UK: Prudential Regulation Authority, 87 p.

Sarowitz, D. (2004). How science makes environmental controversies worse. *Environmental Science & Policy, 7*(5), 385–403.

Soares, M. B., Alexander, M., & Dessai, S. (2017). Sectoral use of climate information in Europe: A synoptic overview. *Climate Services.* http://dx.doi.org/10.1016/j.cliser.2017.06.001.

Twigger-Ross, C., Brooks, K., Papadopoulou, L., & Orr, P. (2015). *Community resilience to climate change: An evidence review.* York: Joseph Rowntree Foundation.

University of Cambridge Institute for Sustainability Leadership (CISL). (2016). *Investing for resilience.* Cambridge: Cambridge Institute for Sustainability Leadership.

Vaughan, C., Buja, L., Kruczkiewicz, A., & Goddard, L. (2016). Identifying research priorities to advance climate services. *Climate Services, 4*, 65–74.

WEF. (2018). *The global risks report 2018* (13th ed.). Geneva: World Economic Forum. Available at http://wef.ch/risks2018.

White, R., & Stirling, A. (2012). Sustaining trajectories towards sustainability: Dynamics and diversity in UK communal growing activities. *Global Environmental Change, 23*, 838–846.

World Bank. (2015). *Investing in urban resilience: Protecting and promoting development in a changing world.* Washington, DC: The World Bank.

World Bank. (2017). *World Population Prospects: Key findings and advance tables.* Washington, DC: The World Bank.

CHAPTER 3

Challenges and Opportunities in Responding to Nexus Shocks

Abstract Nexus shocks are non-linear, spanning multiple sectors and geographies, with decisions often made with a sectoral focus. This can lead to failures to consider the impacts on and interactions of other sectors and stakeholders. A number of challenges and opportunities emerge when examining the impacts of climate shocks to the food, energy, water, environment nexus, particularly when exploring the relationship between society, the system on which it depends and its components (e.g. infrastructure, healthcare etc.). A system's vulnerability and exposure to the risks produced by nexus shocks will affect its capacity to respond and the behaviours of people within it. This is where the co-production of approaches and space for bottom-up initiatives can pave the way to overcome challenges that emerge from nexus shocks and facilitate the design of sustainable and resilient responses to nexus shocks.

Keywords (Mis)communication · Co-production · Collaboration Narratives · Risk · Resilience · Society

© The Author(s) 2019
C. Howarth, *Resilience to Climate Change*,
https://doi.org/10.1007/978-3-319-94691-7_3

HIGHLIGHTS

- Challenges emerging from nexus shocks include the interconnected nature of systems, communication and collaboration, the characteristics of decision-making processes, weak links, over-reliance on resources, social and cultural factors and the nature of responses to nexus shocks.
- Opportunities emerge from nexus shocks which can materialize from challenges: strategic thinking, collaboration and communication, anticipating societal responses, process, identifying key system vulnerabilities, building up capacity.
- There are strong similarities and overlaps in challenges and opportunities identified in the context of climate shocks to the food, energy, water, environment nexus.
- A thorough understanding is needed of the relationship between society, the system on which it depends and its components (e.g. infrastructure, healthcare etc.), and how these are affected by the impacts from shocks, to assess how a nexus shock can affect resilience and capacity to respond.

Climate and weather shocks are characterized by low probability, low frequency of occurrence but with high impact on the resources, stakeholders and processes that exist and interact at the energy, food, water, environment nexus. With this, comes natural and social processes that enrich the complex web that acts to exacerbate or mitigate impacts from these initial shocks. These impacts range from misaligned decision-making processes in the aftermath of a climate shock, timing of responses, lack of communication between local and national stakeholders, and lack of understanding of the social dimension of infrastructure and food, energy, water, environment resources. These are part of an iterative and interwoven process that can act to exacerbate or mitigate the impacts caused by the initial climate shock, be it a heatwave or extreme flooding. These also demonstrate a range of similarities in terms of the prediction of the shock, impacts on and responses of infrastructure, experiences of shocks felt at the local level, how finance and insurance processes interplay, and the role and leadership of governments and governance processes.

Nexus resources are depleting at a rate faster than the natural environment can replenish them yet human activities act to further put pressure on these through over extraction and consumption whilst contributing to global climate change leading to climate shocks. This has led to a global movement of societal responses which acknowledge this underlying challenge, and provide an examination of sectors working at the nexus as entities that interact. These responses demonstrate the inter-linkages between the (energy, food, water, environment) sectors and economic, financial, political and demographic processes that contribute to the complex nature of these shocks. Responses to climate shocks to the nexus are often characterized by being bound to certain sectors with limited diversity of processes reflecting the needs and practices of those external to these sectors which may also be directly or indirectly affected by shocks (Ilin and Varga 2015). These responses are often developed and implemented within certain time frames with less consideration for long-term implications and responses of a system as a whole (Sterman 2012). However, examining climate shocks, their impacts and responses needed through a 'nexus lens', which enables the transition and transformation across sectors and stakeholders (Howarth and Monasterolo 2016: 54), provides a useful approach through which to deliver solutions to climate shocks:

> The nexus allows for a more holistic understanding of (un)intended consequences of policies, technologies and practices whilst highlighting areas of opportunity for further exploration. Nexus thinking represents a multi-dimensional means of scientific enquiry which seeks to describe the complex and non-linear interactions between water, energy and food systems, with the climate, to support understanding of their wider implications for society. (Smajgl et al. 2016)
> The concept of the energy-food-water nexus captures interconnections, dependencies and linkages between production and use of energy, food and water resources. (Howarth and Monasterolo 2016: 54–55)

WHO EXPERIENCES NEXUS SHOCKS?

Nexus shocks affect all individuals and organizations who are impacted directly or indirectly by immediate climate shocks to the nexus at different scales, be it city, regional, national, transboundary or the global level (Dai et al. 2018). The resources within the nexus that frame its lens (notably food, energy, water, environment) characterize the policies and sectors that will be affected (Venghaus and Hake 2018). These range from international and national food policies (e.g. Common Agricultural Policy), energy policies (Renewable Energy Policies), water policies (Water Framework Directive, Nitrates Directive, River Basin Management Plans, Shipping, Irrigation, Water Restrictions) and more general policies (e.g. transboundary, public participation, imports, policy coherence etc.). The sectors most engaged in the nexus and subsequently affected by nexus shocks include agriculture (land, food, farming), energy (fossil energy, renewables, infrastructure, imports, efficiency, consumption, supply), and water (availability, quality). The most prominent stakeholders working on nexus issues include United Nations Agencies, international organizations, research groups, NGOs, private companies, national governments and universities (Endo et al. 2017). When examining the sectoral and resource levels these would include those working at the local, regional, national or international levels and involved in the production, the consumption, the disposal and the design of decisions related to food, energy, water, environment resources. In the UK this would include, but not be limited to, companies that provide water and sewerage services, energy suppliers and distributors, the Consumer Council for Water, regulators, legal and accountancy firms, government departments and agencies (e.g. Department of Environment, Food and Rural Affairs, Department for Business, Energy and Industrial Strategy, Department for Transport, Environment Agency, Food Standards Agency, Natural England, Committee on Climate Change), information and data providers (Met Office, universities, financial institutions), health government department and agencies (Department of Health, Public Health England) and consumers.

The range of stakeholders working across different nexus sectors are characterized by different cultures, behaviours, priorities and processes, all of which will be affected by a nexus shock. The complexities that result from a nexus shock manifest themselves within internal organizational processes of individual and groups of stakeholders and feature in their evolution in response to these shocks. A shock will impact upon decision-making processes regarding the flow and availability of water, energy and food resources which will be most vulnerable to direct impacts (such as damage to crops or increased vulnerability to infrastructure and buildings), or to indirect impacts (such as changes in resource prices, impacts on people's well-being, and health implications from flooded drains). This may have severe and immediate effects on sectors that rely on energy, food and water through damages to crops, from periods of extreme heat or flooding, which will affect production of these resources leading to significant impacts on revenues. These impacts can be devastating, and may also be indirect in terms of their timescale where they may not instantly materialize or be observed for a number of days, weeks or even months after an initial shock. However they can also highlight vulnerabilities of certain resources, stakeholders, individuals or processes thereby shedding light on areas where resilience needs addressing in order to reduce worsening of impacts or alleviate potential future impacts.

Climate extremes are likely to have associated damages and costs potentially tripling by the 2020s and being over ten times current damage (estimated at €3.4 billion per year) by 2100 (Forzieria et al. 2018), with damage from heatwaves likely to rise significantly. Whilst heatwaves are currently rare in the UK, the 2017 Climate Change Risk Assessment predicts that summer heatwaves will become the norm by the 2040s and premature heat-related deaths will increase by a third by the 2050s (Committee on Climate Change 2016). Such severe and prolonged heatwaves have wide impacts within cities across the food, energy, water, environment resources as well as on stakeholders working across the nexus from policy, businesses, transport, health and social care services with effects felt at the local level. Heatwaves exacerbate the heating properties of urban heat islands (De Ridder et al. 2017; Zhao et al. 2018) with large and compact cities, in particular, playing a more exacerbating role on urban climate (Zhou et al. 2017). They are predicted to increase in frequency, intensity and duration across 571 cities in Europe with the number of heatwave days likely to increase in Southern cities, however higher maximum temperatures increases more likely experienced in Central European Cities reaching up 1.5–14 °C particularly in cities that have less infrastructure and populations adapted to extreme heat (Guerreiro et al. 2018). Coupled with the projected growth of population, whereby 50% of global population live in cities projected to reach 70% by 2050 (Heilig 2012) and within the European Union being 75% of population presently projected to grow to 82% by 2050 (UN HABITAT 2011), urban areas face an important challenge of ensuring the resilience of infrastructure, and well-being of their citizens.

The factors that make cities and the people, systems and structures within them vulnerable to high temperatures are complex and dynamic.

These include quality of housing and the built environment, local urban geography, household income, employment, social networks and perceptions of risks (Benzie 2014). This can lead to a range of detrimental impacts such as exacerbating health implications of vulnerable people caused by overheating in buildings, particularly during night times (NHBC Foundation 2012). These all affect people's ability to respond in order to better adapt to these extreme heat conditions. Exposure to such high-temperature extremes over prolonged periods of time depends upon knowledge about how the climate system will change in future, how the population will grow, and the locations of populations in future. The latter two elements are known to increase exposure to heat over the coming decades, although this can occur at different rates in different regions, with further complexities emerging from the non-linear nature of increases in this exposure (Harrington and Otto 2018). With growing population comes increasing productivity and creativity but so do exposure to shocks and growing pressures on infrastructure systems which cities so heavily rely upon (Pregnolato et al. 2016; Adger et al. 2007).

The way in which people experience nexus shocks, and specifically heatwaves, provides a window through which responses to such shock events can be better tailored and targeted. In the UK, individuals tend to demonstrate feelings of nostalgia when thinking about past, hot summers where positive feelings towards these events can lead to the perception that they are safer than they actually are (Finucane et al. 2000). Demonstrating proximity to such events, or increasing salience is a complex process that can lead to intended positive effects, no effects, or can even backfire depending on individual and contextual characteristics (Brügger et al. 2015). For example, when the NHS releases its Heatwave Plan with messages to communicate the risks of extreme heat events, research by Taylor et al. (2017) demonstrated that this can, in fact, trigger positive emotions and hence jeopardize plans and strategies put in place to encourage behaviour change and minimize exposure and vulnerability of individuals. In addition, they show that people tend to perceive flooding events to be more likely to occur than heatwaves; this further jeopardizes mitigation and resilience efforts. Thus, utilizing negative 'affective' experiences of heatwaves, such as demonstrating unpleasant experiences of hot summers (Bruine de Bruin et al. 2016), may be more effective in informing beliefs about climate change and subsequent behaviours to protect against extreme heat (Taylor et al. 2014).

Challenges Arising from Nexus Shocks

Examining nexus shocks provides a lens through which to explore the impacts they subject the food, energy, water, environment nexus to. Simultaneously, they provide insights into the way in which the nexus provides an innovative way to represent the complexities and interactions of nexus resources with the climate and societal activities (Howarth and Monasterolo 2016). Responding to nexus shocks presents both opportunities and challenges (Tables 3.1, 3.2) and central to this is the need for increased understanding of stakeholders' perceptions of climate shocks to the nexus. This will enable 'cross-stakeholder collaboration to fully capture the current, and evolving landscape of decision-making in this space and to develop reflective, robust and resilient approaches that can better capture the realities of these processes when applied in 'live' shock scenarios' (Howarth and Monasterolo 2017: 109). Adopting co-production approaches by which the range of decision-making processes, stakeholder interactions, resource and sectoral interdependencies between stakeholders from across energy, food, water environment sectors are collaborative and iterative, offers space to build on the challenges and opportunities that emerge. Indeed such a transdisciplinary approach relies upon pro-active engagement of these stakeholders from across nexus sectors to share their expertise, skills, resources and lessons learnt in all stages

Table 3.1 Challenges to responding to nexus shocks. From Howarth and Monasterolo (2016)

Communication and collaboration	Evidence production
	Process
	Communication
	Audience needs
	Trust
	Conventional and social media
Decision-making processes	Response
	Responsibility
	Informing and lessons
Social and cultural dimensions	Culture
	Judgement versus evidence
	Disconnection from the issue
Nature of responses to shocks	Uncertainty of science
	Cost of resilience
	Complexities of processes

Table 3.2   Informing responses to nexus shocks. From Howarth and Monasterolo (2017)

Strategic thinking	Precautionary approach
	Considering the bigger picture and context
	Active role assignment and shared responsibility
	Co-production of vision
	Evidence and judgment-based decisions
Collaboration and communication	Creating a common language and cohesive narrative
	Capturing lessons learnt
	Adopting iterative process
	Strategic and co-produced evidence and processes
	Mapping commonalities and differences
	Managing evidence dissemination
Anticipating societal responses	Blending natural and social science insights
	Negotiating a social contract with infrastructure
	Assess cumulative impacts of small societal changes
	Consider response capacity and evidence of responses
	Understand psychosocial responses to risk
	Combining short and long-term thinking
Process	Internal learning and understanding of skills and culture
	Stakeholder humility
	Clear message and messenger
	Combining resource and skills across scales
	Innovative decentralization of decision-making
	Encourage consumer engagement in responses

of knowledge and response development. This provides a clearer picture of the needs, expertise and limitations of those participating in the decision-making process (Howarth and Monasterolo 2016).

Communication and collaboration are particularly important when embedded within decision-making approaches as a process in itself, in the generation of evidence, facilitating knowledge exchange and integration, assessing audience needs, building trust and using conventional and new forms of media (Howarth and Monasterolo 2016). This is fundamental in ensuring the production of evidence and scientific advice that leads to robust decision-making for nexus shocks. This can be a challenge to achieve as, due to the nature of the food, energy, water, environment resources within the nexus, the stakeholders and processes that interact within it, a range of issues can emerge. This can reveal important differences in language, definition and terminology used by different stakeholders across nexus sectors as well as clashes in skills, capacity and resources that can conflict with needs for immediate decision-making processes. In addition, there are multiple types of data, methods, decisions and frameworks which are used (or lost due to historical and turnover issues within organizations) leading to the production of different evidence types and judgement-based decisions which can lead to single-focused approaches to responses and exclusion of external and unfamiliar evidence. The needs of end users and audiences are not always met particularly when evidence available (e.g. providing projections of future climate shocks) or confidentiality issues limiting available evidence, may not match decision-maker's needs. Combining this with different degrees of uncertainty and complexities of the shocks and impacts can further impede decision-making processes to materialize effectively. The process through which evidence is communicated and the mechanisms and media to facilitate this can further create challenges. Timing is one such challenge where academic evidence is often produced on longer timescales (e.g. years) compared to what is required for the formulation of decisions to respond to nexus shocks (e.g. days). This requires closer working between academics, practitioners and policy makers to ensure academic research aims are co-produced with these actors and to ensure they best fit societal needs. Failing to do so can affect trust between stakeholders, and mismanaged information volumes (i.e. information overload), where a lack of communication can lead to siloed working and lack of knowledge exchange and collaboration. The very nature of a climate shock, and its subsequent impacts to the food, energy, water, environment nexus, can create challenges in themselves by affecting communication channels whether they be through traditional (e.g. TV, radio etc.) or newer forms (e.g. social media) of media, or leading to stretched resources available for communication between key actors and the public during a shock.

When a shock occurs, the multiple resources, sectors and stakeholders affected across the nexus can lead to confusion in regards to whose responsibility it is to address the whole or specific aspects of responses required as well as to accept responsibility for decisions made and any implications that may emerge. Local Resilience Fora in the UK, for example, are partnerships made up of multiple agencies and representatives from emergency services, local authorities, the National Health Service and the Environment Agency among others. They plan and prepare for incidents and emergencies experienced at the local level building on the identification of potential risks and designing plans to prevent or mitigate any expected or unexpected impacts. Due to the complex nature of shocks, direct impacts may be observed and experienced in the immediate aftermath of a shock, however indirect, transboundary, slow or emerging impacts may not be immediately apparent. Who is affected by these shocks or responsible for responding to them with this context in mind can lead to challenges in overarching decision-making processes. Accepting responsibility (and associated costs) for developing responses may also be linked to assessments of the positive and negative implications of these responses which may not lie immediately within the confounds of the system which initially experienced the shock. Similarly, timescales can create significant challenges in decision-making processes, particularly when scientific evidence may be developed within different timeframes required by decision-makers to ensure responses are based on sound and up to date evidence. Responses can be further built upon evidence drawing on lessons learnt from historical shocks and how these can be extrapolated to future shocks to ensure pro-active rather than reactive responses are developed.

Nexus shocks are characterized by how they affect food, energy, water, environment resources as well as the web of interactions between stakeholders, sectors and decision-making processes within them. This implies a mix of cultures, behaviours, values and priorities which further impede the responses that occur and their vulnerability to external impacts, particularly when nexus shocks are experienced differently by different sectors and engender different, sometimes conflicting, responses. Understanding of the cultural and social nature of these shocks and responses, how these are interpreted and how they interact to mitigate or exacerbate the initial shock is crucial to ensuring responses align with societal needs and minimize any potential negative impacts. Many decisions in the aftermath of a nexus shock, if there are gaps or delays in expert advice, can rely heavily upon judgment-based decision-making processes (Mach et al. 2017) with a focus on legacy, assessment of available evidence, and preferences for evidence type. This is often predetermined by a need to minimize costs associated with decisions and ensure risks are isolated and addressed, and can indirectly lead to a disconnection from the immediate shock.

As discussed in Chapter 2, communication and collaboration are vital to ensure the most appropriate and robust evidence informs decision-makers at all levels within the context of a nexus shock.

However, this can lead to challenges in informing decision-making. This may produce a lack of clarity as to who is leading a particular response or indeed what that responsibility entails in practice, and what degree of responsibility individual stakeholders and stakeholders have in that regard. The misalignment of timescales between scientists producing evidence which underpins decision-making and the requirements of decision-makers relating to the evidence they need can impede timely responses. The social and cultural dimensions of individuals and organizations that operate within the nexus and are vulnerable to shocks can lead to confusion about different stakeholder needs and resources, potentially resulting in misaligned decisions. In certain shock situations, there may be a challenge around what the most appropriate response should be as there may be an over-reliance on scientific evidence, which can be imprecise, fraught with uncertainties and is constantly evolving, leaving little room for judgment-based decision-making. With this in mind, individual interests may inadvertently be put first, often led by the desire to increase the resilience of assets affected by nexus shocks (or with the potential to be affected by these shocks) in a cost-effective manner. This can at times lead to little consideration for the implications of such processes to others within the nexus (Howarth 2016a).

Opportunities Emerging from Nexus Shocks

The challenges that manifest themselves when a climate shock to the nexus occurs lead to a range of opportunities (Table 3.2) for decision-makers to alter and adapt their own practices to ensure maximum resilience and reduced vulnerability (Howarth 2016a). Employing strategic thinking with a greater emphasis on precautionary approaches as opposed to reactionary ones will enable a more visionary approach to be developed ensuring unanticipated risks, needs of other stakeholders and potential cascading risks are considered. This can result in increased collaboration between academics, policy makers and practitioners who each contribute to evidence which underpins decision-making processes and the formulation of these processes. By encouraging and undertaking greater engagement with different stakeholders, sectors and resources, processes of co-production can enable better-designed aims reflective of reality, a deeper understanding of the way in which each operates in a nexus shock and non-shock scenario, and a closer alignment with the needs of different end-users. This can enable more effective working practices with other stakeholders to understand how societal and public responses to nexus shocks manifest themselves and evolve with better anticipation of how these may make initial impacts and responses worse leading to decision-making processes which are more challenging to design, implement, monitor and evaluate. A re-negotiation of existing 'social contracts' can then occur between society, assets, infrastructure and with stakeholders involved in the nexus, and affected by climate shocks. This would enable ordinary civilians to play a more active role in shaping responses to nexus shocks and ensure decision-making processes more accurately reflect the needs and response processes that the public undertake. This application of a holistic or systems approach to managing nexus shocks will enable a broader understanding of the implications of decisions made on other stakeholders and sectors. In addition, this enables decision-makers to better understand the context within which their decisions are made and how these evolve over time, adopting an iterative approach open to flexible management of this process.

By their nature, nexus shocks require strategic thinking to ensure robust decisions are made to effectively respond to their impacts. Often, precautionary approaches are adopted, based on evidence, historical lessons and experiences, that enable better foresight and projections of impacts, needs, costs and processes to be adapted to the evolving context within which a shock occurs. This facilitates processes by which resources can be deployed to address specific risks that may affect food, energy, water, environment resources, their social dimensions, and the infrastructure which enables their growth, transport and development. With this, and transparency around roles, leadership and trusted structures, strategic thinking can ensure consideration for potential future, cascading, emergent impacts and feedback effects which can further increase system and societal vulnerabilities. Co-production can play a central role here by incorporating evidence-based and judgment-based decisions from across the range of actors and sectors that hold a stake in both impacts and responses to nexus shocks. In so doing, it enables a larger vision to be adopted with clarity on ownership of impacts and risks, responsibilities at the community, local, regional and national levels and mechanisms for knowledge exchange within this complex system.

Whilst identified as a barrier to decision-making, collaboration and communication are also seen as an opportunity to strengthen relationships, knowledge exchange and decision-making processes. If communication and collaboration are seen as a barrier, if their sources, nature and impacts are better understood then common languages and a joint vision between stakeholders can be developed to establish mechanisms through which to communicate effectively and resources pooled to target where needed. This vision can ensure better strategizing to filter and prioritize evidence to inform decision-making, co-produce evidence where needed and target this to specific audiences based on the context, impacts experienced, needs and resources available. Adopting an iterative and reflective process, whereby systems in place and the stakeholders working within them actively seek to self-evaluate their progress and work and embed learnings into their decision-making processes, can ensure a smoother flow of information to underpin effective communication. This can also help explore the range of scenarios of impacts, responses and vulnerabilities that could ensue from a particular shock. Feeding this into decision-making processes can increase the credibility of responses initiated, facilitate and better manage the flow of evidence produced and used, improve trusted relationships between stakeholders and limit costs and risks of the initial shock to a system. The utilization of new and innovative methods can be an effective way to do so in order to ensure timely and instant delivery of evidence whilst accounting for potential risks of mismanagement or overuse. This can thereby contribute to building a powerful narrative on how nexus shocks impact food, energy, water, environment resources and stakeholders, demonstrating the benefits of collaboration, communication and knowledge exchange.

Nexus shocks impact different aspects of society in varying ways and hence decision-making should consider and anticipate the social responses that will emerge, as failing to do so may mean inadequate decision-making processes, misaligned with societal needs, are put into place. It is important to fully understand the relationship between society, the system on which it depends and its components (e.g. infrastructure, healthcare etc.), and how these are affected by the impacts from shocks. A system's vulnerability and exposure to the risks produced by nexus shocks will affect its capacity to respond and the behaviours of people within this. For example, people may be unwilling to leave their homes during flooding or heatwave events, when evacuation processes are underway, further endangering them and undermining strategic approaches in place. Overcoming this may require a suite of initiatives that embed short and long-term thinking, including the democratization of the process with public and cross-stakeholder involvement, bottom-up approaches and the incorporation of lessons learned across sectors and scales.

A sound process is therefore needed and can emerge following a nexus shock enabling more efficient governance structures in place to address impacts of nexus shocks. In an ideal scenario this would be sufficiently resilient to enable flexibility and iteration with planning processes, fluid mechanisms and structures to ensure rapid responses where needed, and decentralization of response development to ensure more tailored and targeted approaches. As discussed above, responses to nexus shocks are complex and depend on the unique characteristics of the shock, its impacts and responses across sectors and stakeholders. From this emerges the requirement to ensure an acknowledgement of current skills and capacity as well as limitations that may exist to embed a broader culture of learning and self-reflection to occur, enabling a clearer understanding of the challenges faced. With this in mind, the resilient and efficient nature of these processes enables flexible planning and more robust decisions to be made that are complementary, rather than contradicting existing routes, products or services in place. Innovative approaches can thus be co-produced by multiple stakeholders, each of whom will benefit from personally investing in sharing concerns, lessons learned and successes to better formulate ways to adapt to and build on nexus shocks.

INFORMING RESPONSES TO SHOCKS

Current responses to climate shocks to the food, energy, water, environment nexus vary according to the specific shock, its characteristics, who and what it impacts, whether it leads to cascading effects, if it has both immediate and delayed impacts, the scale of the shock and its impacts, the nature of responses, whether such a shock has occurred in the past and decision-making processes already be in place. Decisions occur at different scales and when designed and implemented at the national level, for example, may lead to a failure to consider implications (negative and positive) for local or even international levels. Decision-makers at the local (for example through Local Resilience Fora) and national levels have a tendency to focus on short-term and sector-bounded problems and associated benefits which places less emphasis on long-term implications for the system as a whole and how short-term responses may exacerbate or mitigate future responses. Decisions made at the sector level, however, can lack diverse participatory mechanisms of different stakeholders affected directly and indirectly by a shock. This can lead to a lack of consideration for other processes in place in other sectors that may have unexpected and unanticipated impacts across other sectors (Howarth 2016b). There may also be rebound, cascading or other negative effects from nexus shocks and decisions that are formulated in response, mainly due to these being characterized by mutual interdependencies with, at times, unforeseeable implications on decision-making processes. Consequently, evidence produced to inform decision-making may not always be effective or may not be shared appropriately between scientists, decision-makers and practitioners when co-producing effective responses and building resilience to climate shocks.

In order to build on the challenges and opportunities identified in this Chapter and ensure these feed into a process of knowledge exchange, capacity building and lessons learned, a clear understanding is needed of how nexus shocks impact stakeholders (and how these stakeholders perceived these impacts) as well as the opportunities that are identified which could be used as springboards for resilience building. Collaboration across sectors and stakeholders is thus necessary to deliver this whilst capturing current and future settings for decision-making that best reflect the real needs and abilities of those involved in the impacts and responses to nexus shocks. In so doing, adopting these will ensure processes that are characterized by transparency, accessibility, engagement with non-academic audiences and main nexus stakeholders, understanding the role of incentives to inform the introduction of climate resilient policies along the nexus ultimately strengthening and ensuring the sustained resilience and embedded resilience culture to shocks in future. Nexus shocks are non-linear, spanning multiple sectors and geographies, with decisions, often made with a sectoral focus, which can fail to consider the impacts on and interactions of other sectors and stakeholders. This is where the co-production of approaches and bottom-up initiatives can pave the way to overcome challenges that emerge from nexus shocks and facilitate the design of sustainable and resilient responses to nexus shocks.

REFERENCES

Adger, W. N., Agrawala, S., Mirza, M. M. Q., Conde, C., O'Brien, K., Pulhin, J., et al. (2007). Assessment of adaptation practices, options, constraints and capacity. In *Climate change 2007: Impacts, adaptation and vulnerability* (pp. 717–743). Contribution of Working Group II to the Fourth Assessment Report of the Intergovernmental Panel on Climate Change. Cambridge, UK: Cambridge University Press.

Benzie, M. (2014). Social justice and adaptation in the UK. *Ecology and Society, 19*(1), 39.

Brügger, A. R., Dessai, S. X., Devine-Wright, P., Morton, T. A., & Pidgeon, N. F. (2015). Psychological responses to the proximity of climate change. *Nature Climate Change, 5,* 1031–1037.

Bruine de Bruin, W., Lefevre, C. E., Taylor, A. L., Dessai, S., Fischhoff, B., & Kovats, S. (2016). Promoting protection against a threat that evokes positive affect: The case of heatwaves in the UK. *Journal of Experimental Psychology: Applied, 22*(3), 261–271.

Committee on Climate Change. (2016). *UK climate change risk assessment 2017 synthesis report: Priorities for the next five years.* Accessed: https://www.theccc. org.uk/wp-content/uploads/2016/07/UK-CCRA-2017-Synthesis-Report-Committee-on-Climate-Change.pdf.

Dai, J., Wu, S., Han, G., Weinberg, J., Xie, X., Wu, X., et al. (2018). Water-energy nexus: A review of methods and tools for macro-assessment. *Applied Energy, 210,* 393–408.

De Ridder, K., Maihen, B., Lauwaet, D., Daglis, I. A., Keramitsoglou, I., Kourtidis, K., et al. (2017). Urban heat island intensification during hot spells—The case of Paris during the summer of 2003. *Urban Science, 1*(1). https://doi.org/10.3390/urbansci1010003.

Endo, A., Tsurita, I., Burnett, K., & Orencio, P. M. (2017). A review of the current state of research on the water, energy, and food nexus. *Journal of Hydrology: Regional Studies, 11,* 20–30.

Finucane, M. L., Alhakami, A., Slovic, P., & Johnson, S. M. (2000). The affect heuristic in judgments of risks and benefits. *Journal of Behavioral Decision Making, 13*(1), 1–17.

Forzieria, G., Bianchi, A., Silva, F. B., Herrera, M. A. M., Leblois, A., Lavalle, C., et al. (2018). Escalating impacts of climate extremes on critical infrastructures in Europe. *Global Environmental Change, 48,* 97–107.

Guerreiro, S. B., Dawson, R. J., Kilsby, C., Lewis, E., & Ford, A. (2018). Future heat-waves, droughts and floods in 571 European cities. *Environmental Research Letters, 13*(3). https://doi.org/10.1088/1748-9326/aaaad3.

Harrington, L. J., & Otto, F. E. L. (2018). Changing population dynamics and uneven temperature emergence combine to exacerbate regional exposure to heat extremes under 1.5 °C and 2 °C of warming. *Environmental Research Letters, 13*(3). https://doi.org/10.1088/1748-9326/aaaa99.

Heilig, G. K. (2012). *World urbanization prospects: The 2011 revision.* New York: United Nations, Department of Economic and Social Affairs (DESA), Population Division, Population Estimates and Projections Section.

Howarth, C. (2016a). *Informing decision-making in response to nexus shocks.* LWEC PP Note.

Howarth, C. (2016b). *Responding to extreme weather events* (ESRC Evidence Briefing).

Howarth, C., & Monasterolo, I. (2016). Understanding barriers to decision-making in the UK energy-food-water nexus: The added value of interdisciplinary approaches. *Environmental Science & Policy, 61,* 53–60.

Howarth, C., & Monasterolo, I. (2017). Opportunities for knowledge co-production across the energy-food-water nexus: Making interdisciplinary approaches work for better climate decision-making. *Environmental Science & Policy, 75,* 103–110.

Ilin, T., & Varga, L. (2015). The uncertainty of systemic risk. *Risk Management, 17,* 240–275.

Mach, K. J., Mastrandrea, M. D., Freeman, P. T., & Field, C. B. (2017). Unleashing expert judgment in assessment. *Global Environmental Change, 44,* 1–14.

NHBC Foundation. (2012). *Overheating in new homes: A review of the evidence.* NHBC Foundation and Zero Carbon Hub. Available at: http://www.zero-carbonhub.org/sites/default/files/resources/reports/Overheating_in_New_Homes-A_review_of_the_evidence_NF46.pdf.

Pregnolato, M., Ford, A., Robson, C., Glenis, V., Barr, S., & Dawson, R. (2016). Assessing urban strategies for reducing the impacts of extreme weather on infrastructure networks. *Royal Society Open Science, 3,* 160023. http://dx.doi.org/10.1098/rsos.160023.

Smajgl, A., Ward, J., & Pluschke, L. (2016). The water–food–energy nexus—Realising a new paradigm. *Journal of Hydrology, 533,* 533–540.

Sterman, J. D. (2012). Sustaining sustainability: Creating a systems science in a fragmented academy and polarized world. In M. Weinstein & R. E. Turner (Eds.), *Sustainability science: The emerging paradigm and the urban environment* (pp. 21–58). New York: Springer.

Taylor, A., Bruine de Bruin, W., & Dessai, S. (2014). Climate change beliefs and perceptions of weather-related changes in the United Kingdom. *Risk Analysis, 34*(11). https://doi.org/10.1111/risa.12234.

Taylor, A., Dessai, S., & Bruine de Bruin, W. (2017). Public priorities and expectation of climate change impacts in the United Kingdom. *Journal of Risk Research.* ISSN: 1366-9877.

UN HABITAT. (2011). *Cities and climate change: Global report on human settlements 2011.*

Venghaus, S., & Hake, J.-F. (2018). Nexus thinking in current EU policies—The interdependencies among food, energy and water resources. *Environmental Science & Policy*. https://doi.org/10.1016/j.envsci.2017.12.014.

Zhao, L., Oppenheimer, M., Zhu, Q., Baldwin, J. W., Ebi, K. L., Bou-Zeid, E., et al. (2018). Interactions between urban heat islands and heat waves. *Environmental Research Letters, 13*(3). https://doi.org/10.1088/1748-9326/aa9f73.

Zhou, B., Rybski, D., & Kropp, J. P. (2017). The role of city size and urban form in the surface urban heat island. *Nature Scientific Reports, 7*, 4791. https://doi.org/10.1038/s41598-017-04242-2.

CHAPTER 4

The Importance of Communication, Collaboration and Co-production

Candice Howarth, Sian Morse-Jones

Abstract Building resilient responses to nexus shocks requires effective communication and collaboration across sectors and stakeholders, yet this is not always achieved. The Nexus Shocks project examined how communication and collaboration could be enhanced, adopting a co-production methodology with policy, practitioner and scientific communities. This chapter discusses the barriers and challenges to communication and collaboration on specific nexus shocks, such as heatwaves and flooding, and identifies pathways to strengthen responses. Co-production provides a constructive way to deliver more salient decision-making processes which incorporate the needs of those affected in managing and responding to nexus shocks.

Keywords Communication · Collaboration · Co-production Resilience

HIGHLIGHTS

- Increased resilience to nexus shocks such as heatwaves and flooding requires effective communication and collaboration.
- Better communication can be achieved by contextualizing warnings; providing digestible, targeted, clear and consistent messaging; using relatable language; and ensuring effective dissemination structures are in place.

© The Author(s) 2019
C. Howarth, *Resilience to Climate Change*,
https://doi.org/10.1007/978-3-319-94691-7_4

- Good collaboration requires: time, trusted relationships, collaborators as *partners*, and joined-up, agile processes, aligned with the needs of those affected.
- Co-production provides a constructive way to deliver more salient decision-making processes which better incorporate user needs.

Uncertainties about climate change are complex and vary according to the context within which knowledge is acquired to inform decision-making, the messenger who is communicating, the objective of this dissemination and the content of the information. Climate change and the risks associated, is challenging to communicate to non-experts particularly over different timescales and geographic locations. People interpret and use information based on values and beliefs with an alignment of these not necessarily resulting in predicted behaviours. These risks are understood in numerous ways depending on how people value the outcome of the risks in question and the way in which these will impact them whether it be economically, environmentally or other (Pidgeon and Fischhoff 2011). Assessment of climate change risks, and understanding of this, can be enriched when specific local level information (i.e. district level) on vulnerability and resilience to the effects of these risks is included enabling more informed decision-making (Bennett et al. 2014).

Predominant issues around communication of climate change rest with an ever evolving science and corresponding economic, social, cultural and political context which leads to a complicated language used to transfer information. Constant interaction between scientists and policy makers is therefore required to ensure decision-making (Bidwell et al. 2013) reflects the most up to date scientific evidence. However, there are numerous issues associated with this: building strong trusted relationships takes time as well as skills, and an appetite for relationship-building which may be lacking; decision-makers will need to consider numerous other factors in their decisions in addition to climate, and the dialogue is not linear containing many complexities. Individual's decisions rest on a web of values, beliefs, preconceptions and more, consequently, different strategies are required to address these various values, beliefs, motivations, perceptions, attitudes and stakes involved. Communication processes should also be based on a sound understanding of social learning, 'the movement of information and practices through knowledge networks, whose structures influence both the pace and qualities of learning as the networks themselves evolve' (Bidwell et al. 2013: 610).

There are a number of reasons why various stakeholders (e.g. policy makers, technocrats, practitioners, decision-makers, academics, those leading responses, the public) are unable to fully engage with climate change and the shocks that emerge. These include: it lackings visibility; it being hard to quantify the local impacts it will have; scientists being unable to predict these impacts with complete accuracy nor what solutions would be the most impactful; if people are able to get to grips with the degree of uncertainty, the level of complexity is ever more challenging; the risks of climate change are for many 'virtual' rather than real: geographically and temporarily distant (Nerlich et al. 2010). There is therefore a need to appeal to people's understanding, emotions and behaviours whilst addressing issues such as engaging, interesting and relevant messaging which considers people's self-identity: 'those desiring change need to engage with people's important values and sources of identity, rather than merely appealing to their short term interests' (Nerlich et al. 2010: 100). Climate change communication is better known, understood and more relevant to different audiences, across different media (e.g. press, film, literature, theatre, policy etc.) yet emissions continue to rise, impacts of climate shocks on the food, energy, water, environment nexus are more prominent and resilience to these shocks remains in a state of progress. This leads to a questioning of the efficacy of what is communicated, the messenger and the ability of the audience to translate their acquired awareness into action. Targeting of climate communication strategies is therefore encouraged with a broader approach inclusive of audiences' stated preferences on what is likely to change their behaviour (i.e. information content) incorporated more readily to deliver impactful change (Hine et al. 2014).

The food–energy–water–environment nexus is characterized by a range of complexities that span multiple sectors that work across this nexus and the resources on which they depend. A nexus approach, which provides insights into the impacts of extreme weather and climate-related shocks, such as heatwaves and flooding, to energy, food, water, environment resources in an interdisciplinary and collaborative way, can help encourage effective decisions and responses by sharing knowledge, skills, expertise, best practices and lessons learned. The frequency and intensity of nexus shocks are predicted to increase both globally and in the UK (ASC 2016), and, being characterized as unpredictable and uncertain with a web of complex impacts, effective responses will require communication and collaboration across different levels of society, governments and the scientific community.

Whilst, clear and effective communication of evidence, and impacts, to specific audiences can raise resilience, this is not always adequately achieved due to a linear process of dissemination (the information deficit model) where access to information is assumed to lead to a mobilization of action. For example, climate change science is often communicated to decision-makers with assumptions that this will adequately inform decision-making processes, yet due to the complexity and uncertainties of climate science, the natural and social dimensions of impacts and responses, and especially where decisions are highly context-dependent, this is rarely the case (Howarth and Painter 2016). Instead, a better alignment is required between what end-users and those working on the ground need and what scientists perceive to be useful (Lemos et al. 2012). These communication challenges must be faced to minimize issues of miscommunication and assess the extent to which future societal needs fit into decision-making processes. Doing so by consulting historical experience would further contribute to overcoming the lack of community understanding of risk, people's desensitization by information networks, and lack of trust in decision-makers and those informing them (Howarth and Monasterolo 2016). However, there are a number of barriers which affect access to evidence which prevents decision-makers from using this to shape responses.

The widespread assumption that providing more information will lead to desired change is flawed. Instead, an emphasis on devising and implementing solutions illustrates the need for mechanisms to facilitate decision-making processes. Climate services for example which are co-developed with end-users (Christel et al. 2017), are useful ways of delivering information directly aimed at apprising decision-makers (Webber 2017), they provide appropriate representation of related governance processes (Clarke et al. 2013), and address potential issues of malpractice (Scandelius and Cohen 2016). In doing so, collaborative processes, built on humans' natural ability to collaborate for mutual gain (Newton 2017), actively pursue mechanisms for conflict resolution (Löhr et al. 2017) and build on and acknowledge the need for more 'complex forms of agent interactions in the production, framing, communication and use of climate knowledge; and in particular, explicit procedures able to tackle difficult normative questions regarding assessment of solutions and the allocation of individual and collective responsibilities' (Tàbara et al. 2017: 31). Beliefs among stakeholders and formal power structures (as opposed to structures of power that are perceived) play an important role in decision-making processes for responses to climate change (such as climate mitigation) whilst prove limited in the implementation of these responses (Ingold and Fischer 2014).

Communication and collaboration are key to ensure resilience to nexus shocks is built (Howarth and Monasterolo 2016) yet deeper understanding of the barriers that exist is needed to ensure decision-making processes align with the needs of end-users and those working across the nexus, without the risk of neglecting certain impacts or stakeholders. Through interviews, the Nexus Shocks project examined how communication and collaboration on nexus shocks, such as heatwaves and flooding, could be enhanced, adopting a co-production methodology with policy (PO), practitioner (PR) and academic/scientific (AC) communities. Key components of communication activities require strategic approaches encapsulating the needs of audiences and stakeholders engaged, the context of evidence and advice communicated, the messenger used to communicate, understanding of how information and advice is received and acted upon, acknowledgement of external and contextual influences and considerations for future implications of current response or non-response.

Communication and Building Resilience to Nexus Shocks

The Nexus Shocks project sought to ascertain how evidence used to inform decision-making in the context of nexus shocks, could be better communicated to more effectively inform decisions. The evidence used by stakeholders ranges from climate or meteorological data, broader environmental data (e.g. hydrological, ecological, geological etc.), social and economic data, evidence from policy, practitioners, utility companies, satellite to more informal, anecdotal evidence. As is evident elsewhere, effective communication of this evidence is hampered by a linear process of dissemination. Examples of positive communications on nexus shocks include the use of captivating, compelling stories and narratives as well as case studies of specific risks and the impacts and solutions that emerged (e.g. for heatwaves impacts and flood impact). The adequacy of communication to inform decisions depends highly on the evidence produced or available, who it is designed for, and for what specific purpose, whether this be preparation, pre- or post-event response. Generally, communication of evidence from experts and scientists is considered to be of a good standard, useful and joined up, particularly for operational responses yet this is less the case for preparation and building resilience. When it comes to providing information to the public and decision-makers this is not considered to be as effective by those interviewed in the project, especially when communicating risk and uncertainty where it was felt the objective often tends to focus more on avoiding immediate panic.

> I think amongst the professional community I think that people are aware of the evidence and understand what's been said and the risk barriers and everything else. That's flood professionals I would say. I think outside that area, not so well. In terms of the public, I think that we haven't found a very useful way of explaining to people they're at risk yet. (PO3)

Different challenges manifest when communicating different nexus shocks, such as heatwaves and flooding. When it comes to flooding, different or bespoke responses may be required depending on the context and specific impacts of the incident, therefore longer-term preparations are needed which align better with the needs of professionals, practitioners and decision-makers than members of the public. According to the interviewees, professionals in the flood community are aware and understand evidence and risks of flooding however there is no clear, useful way of explaining flood risk to the public without referring to frequencies, probabilities, flood return periods and statistics which results in confusing non-expert audiences. To the public these often represent abstract measures, with a 1 in 200-year event assumed to happen only once every 200 years, rather than understanding that this risk could occur in consecutive years, curtailing the ability to make informed decisions based on accurate understanding of risks. In relation to heatwaves, whilst instructions and solutions to manage public responses are salient and simple to communicate (i.e. keep cool, drink enough etc.), there is concern that awareness of heatwaves risks is low. Culturally, complexities exist around communicating heat risks, which is more challenging when people enjoy the sunshine, and there is no physical disruption, meaning that the public may fail to adequately recognize the serious risks associated with exposure to prolonged heat events, especially for vulnerable people (such as above a certain age, having health issues and less able to move) and the simple solutions outlined earlier may not be followed. For example, during the European 2003 summer heatwave, 2000 heat-related deaths were recorded in the UK, despite various television, internet and newspaper communications informing the public how to cope with the heat (Met Office 2015). Generally, there are concerns that insufficient information is given to the public, and that when this is provided it does not use appropriate language to mobilize action. There are also concerns that too much information can lead to people feeling overloaded, not knowing how to respond to information provided leading to a general disengagement. The communication of risk and uncertainty is considered poor, with the UK seen to be lagging behind other countries due to a lack of understanding of flooding and heatwave risks, misperceptions towards these shocks and inappropriate use of technical terminology in public communications. Return periods, for example, are known to be a problem as they don't explain impact or size and anecdotal evidence can help more effectively increase salience to likelihood and probability language.

When providing advice, scientists and academics are not always packaging their findings in ways that can successfully reach practitioners and decision-makers. Whilst efforts are underway to improve this, there is insufficient focus on the language used, for example, communication on nexus shocks such as heatwaves and flooding needs to be improved to better reflect the need to move away from complex language (e.g. probabilities) that may mean little to non-expert audiences. There is a general sense that communications have often failed to raise awareness and engender a sense of ownership and responsibility towards the appropriate management of risks, particularly among the public. This is mirrored by difficulties in maintaining interest and engagement levels in between occurrence of shock events which results in reactive (rather than proactive) responses. In order to address this, communication of this nature could be improved by including end-users earlier in the production process by adopting a co-production approach (explored later in the chapter).

> I think the preparedness and resilience, so the pre-event stuff, I think that's probably where the biggest improvements in communication of evidence could be made. And it's just the normal challenges. You know, a huge amount of information is generated by the academic community and international organisations and others and it's just not communicated to decision-makers in the right way. (PO2)

A range of options exist to improve the communication of nexus shocks to strengthen decision-making processes (see Table 4.1). Firstly, contextualization of weather and climate warnings in relation to recent events, local surroundings, and situational contexts can help people understand the scale of what they are experiencing, to trigger action to reduce impacts and fine-tune information based on the local implications and solutions required. The mechanism adopted for communicating messages needs to be relatable and can be strengthened by using first-hand experiences; stories and narratives that resonate with policymakers, enabling them to understand why support is needed. Secondly, digestible information, targeted to audiences with a clear purpose of the message is needed. A reduction in the detail of information is widely encouraged by

Table 4.1 Options for improving communication

Option	Rationale	Examples of tools
Contextualization of weather and climate warning	To increase understanding of the scale of the shock, to trigger action to reduce impacts and fine-tune information based on the local implications and solutions required	Stories and narratives Comparisons to recent events Communicating long-term issues in the context of present day Use of situational warnings Maps to locate risks and impacts
Digestible and targeted information	To enable audiences to quickly establish the salience and relevance of risks and required responses, enabling efficient filtering and triggering appropriate action when needed	Simple, clear messaging Thumbnail graphics of location of risk Signposting to more detailed information
Appropriate use of language	To enable audience to properly understand and engage with messages, to trigger intended response	Relatable, everyday language appropriate to audience Clarifying meaning of words or phrases to different collaborators
Adequate structures for communication	To ensure correct communication feeds through to the intended audiences at the right time	Range of communication channels e.g. local radio, newspaper, social media etc Harnessing mobile communications for live data Virtual interactive common operating picture platforms

Nexus Shocks participants whilst ensuring in-depth information, where needed, provides a clear set of warnings and alerts, better signposting, and enables efficient filtering of alerts. Value is placed on the use of a single voice, single message coordinated through collaboration with other stakeholders to deliver a clear, consistent message aligned with others, which is more effective and trusted by end-users. Thirdly, the use of language is particularly important in communication and framing. Terms such as 'climate change' and 'nexus' have been found to hold little salience with some audiences, yet talking in terms of 'hot weather' is immediately relatable. This raises the challenge of the range of languages and lexicons used in different organizations and the need to demonstrate credibility and trust, without which action can be undermined. Fourth, structures must be in place to share and disseminate effectively through a range of communication channels, for example, trusted word-of-mouth, TV, newspapers, local radio, and social media, ensuring correct information is feeding through planning, preparation, and response stages of a shock. Clarity is needed on the expected impact and action required; this increases relevance and engagement, and leads to better aligned behavioural responses. This can be more effectively done by working alongside the growth in technology for communication during emergency events and building on strong relationships with partners who have access to evidence.

The Role of Collaboration in Increasing Effective Responses

Whilst communication enables constructive dialogue to occur to design nexus shock responses, how this leads to increased resilience and stronger decision-making processes depends on how the stakeholders involved are able to work together. When a nexus shock occurs, those involved and affected will undoubtedly end up engaging in a process by which they depend upon each other for access to evidence, knowledge and expertise and share experiences of the impacts of the shock and historical lessons on mitigating these impacts. This is not without its gaps, and there is a need to understand how collaboration can be harnessed to greater effect to improve decision-making and resilience to nexus shocks.

Collaboration is seen by Nexus Shocks participants as vital to ensure the most robust evidence informs decision-making and to design and implement appropriate responses. The type (and length) of collaboration following a shock event depends on: its nature; whether it manifests as a single event (e.g. flash flood) or over a long period of time slowly growing in intensity and impact (e.g. heatwave); the current developments that can affect its impacts and responses; the stakeholders involved; and the nexus sector(s) affected. A shock provides an opportunity to bring stakeholders together to target specific responses in the immediate aftermath and to plan pro-active responses for future shocks. Collaboration between nexus stakeholders enables a more comprehensive view of the shock at hand, captures the needs and objectives of stakeholders affected by shocks and leading responses, and enables a rapid assessment of potential impacts to infrastructure, human lives, economy, culture and the environment. In doing so, it facilitates the exchange of information and better access to data, evidence or expertise providing a rich picture of what is happening on the ground. In so doing, it enables a hands-on view of how a system works and brings together practical expertise and theoretical knowledge, enhancing the quality of data, relationships and responses.

Locally, collaboration increases effectiveness in managing and responding to flooding or heatwaves particularly where communities are more resilient and better able to support those most vulnerable. Nationally, collaboration between government departments, utility companies, NGOs, or community groups can lead to a better assessment of the needs and capabilities of those actors ensuring responses are more aligned with the needs of those impacted. It enables a pooling of resources and sharing of work, providing access to expertise that may not necessarily be available in-house. The process of collaboration enables a better and clearer understanding of issues and knowledge gaps and ensures research is framed around needs-based questions thereby maximizing the potential for responses to be more adequately formulated.

> The main benefits? I think different organisation have different responsibilities to different things and it is only by working together that holistically there is a collective response. If there wasn't that collaboration I am sure there would be things that would slip through the gaps in the response. But by working together and then having a structure to respond together to any issues, I think it helps minimize any impact of that. (PR6)

However, collaboration processes are not without their challenges. Top down engagement often fails to work in the long-term due to a lack of incorporation of the wider knowledge-base, including those operationalizing responses on the ground. This can lead to bad practice when a single collaborator is left to conduct post-shock work as other collaborators are too under-resourced to own responsibility of the process. For example, situations may arise where the longer-term recovery (and preparation) processes rest predominantly with flood-affected communities. The nature of the collaboration process means that at times organizations may not be able to speak directly to end-users as it may be difficult to assess whether people representing organizations have traction at decision-making levels. Different organizations and individuals may have fundamentally different methods of working, where some may be more inclined to collaborate, sharing as they go along, whereas others have a different culture. This means that collaboration can be hard to carry out and sustain, and at times may not be appropriate when involving a wider group working to different priorities and cultural settings. Apathy and unwillingness to engage requires a degree of expectation management, for example, in the context of a flooding incident, communities may expect local councils to demonstrate leadership and response rather than taking ownership themselves. Collaboration takes time (and resources), particularly when adopting a co-production approach (as we will explore in the next section) which presents barriers to actors who are time poor and is further exacerbated by limited funding or the lack of structures to deliver processes at the appropriate scale.

Collaboration can be strengthened to support better decision-making in relation to nexus shocks in a number of ways (see Table 4.2). Good collaboration depends on relationship enabling processes built on trust where communication channels are well established. The creation of networks enables inclusive cross-stakeholder collaboration facilitating knowledge exchange on key responses to weather and climate extremes whilst building capacity and knowledge. These networks enable a deeper applied approach to formulate and implement responses, and can bypass delays faced at policy or leadership levels, which may be overloaded by cumbersome bureaucracy, where decision-making processes are far removed from the realities of a nexus shock on the ground. For example, the London Climate Change Partnership (LCCP), which is made up of public, private and community sector organizations, works within and across sectors to increase resilience in London to extreme weather-related events and climate change, undertaking a range of activities. It coordinates the 'Heat Risk in London' group which facilitates knowledge exchange for planning, preparing and responding to heatwaves risk,

Table 4.2 Options for enhancing collaboration

Option	Rationale	Examples of tools
Relationship enabling processes, over time	To build trust and understanding, to create strong, supportive relationships, enabling access to data, knowledge, and expertise, and efficient and effective responses	Invest time and effort to understand language, drivers, skillsets, personalities, needs, priorities, capabilities and ways of working, before shock events Use of established communication channels Creation of networks Having agreements in place to enable easy sharing of data
Active collaborators as *partners*	To facilitate more joined-up, comprehensive and coordinated responses with a deeper understanding of needs and greater ability to influence	Doing *with* people, rather than to people Aligning priorities and timescales Pooling resources, sharing work Policymakers/practitioners partnering with academic research
Flexible, agile processes,	To facilitate proactive, preemptive and rapid responses, aligned with the needs of those affected	Bringing stakeholders together immediately after an event to target specific responses, and to plan for future shocks Applying big picture, long-term thinking, working backwards from desired outcomes Crowdsourced approaches for finding solutions to issues Establishing processes to maintain and make use of historical and institutional memory on nexus shocks Knowledge exchange grants Flexibility to explore different options 'outside the box' through open funding calls
Inclusive and aligned with needs	To ensure the needs and objectives of sectors and stakeholders, including those affected by shocks, are considered, enabling more effective responses, with greater reach	Drawing on a wide range of expertise, stakeholders, sectors, organizations and individuals Enabling and providing opportunities for stakeholders to collaborate, steer and influence Inviting community representatives onto decision-making panels and boards Funding to support cross-sector and multi and inter-disciplinary responses

convening discussions, sharing case studies and learning, and engaging academic experts; and through its Observing London project is working with organizations to create a network of weather stations to improve the quality and accessibility of weather data at the microclimate (LCCP 2018a, b). More active collaborations such as these, are needed between policymakers, practitioners and academics as *partners* rather than stakeholders to facilitate more joined up approaches which provide greater ability to influence. This is further strengthened by adopting flexible and agile processes where contributors benefit from innovative ways of working and sharing in a way that aligns with the needs of affected communities. This enables a stronger focus on working backwards from the desired outcome, taking pro-active and pre-emptive approaches for shock events, with the bigger and longer-term picture in mind. Collaborations can benefit from stronger emphasis on maintaining historical and institutional memory to ensure that when a shock re-occurs (as they are projected to do), relevant knowledge and expertise is not lost. Ultimately, responses to climate risks are under-funded and funding mechanisms are needed whereby academic, policy and practitioners are eligible and have more leverage over how money is spent to ensure findings can be applied.

CO-PRODUCING RESPONSES TO NEXUS SHOCKS

Innovative approaches that adopt a transdisciplinary lens, by engaging with academics and key stakeholders who will be affected by nexus shocks and have an interest in work being carried out, are increasingly being used to address these important and growing societal challenges (Bammer 2013). Co-production is one such type of approach which enables a more efficient facilitation and navigation of the relationships and tradeoffs between energy, food and water resources and the actors involved whilst subverting any possible challenges that may arise (Howarth and Monasterolo 2017; Vesselinov and Zhang 2016). Co-production has been applied across a range of fields and contexts (e.g. climate change: Howarth et al. 2017; urban development: Omondi et al. 2014, multi-level governance and planning: Watson 2013, community engagement, political ecology, public policy) and is characterized by a focus on end-users (Voorberg et al. 2014), incorporation of non-academic actors to design its aim (Polk 2015), being inclusive, collaborative, integrative, self-reflective, useful and embracing challenges (or 'uncomfortable moments') within its development, design and implementation process. In essence, it enables a democratization of politics (Jasanoff 2010) and decision-making processes ensuring a large, diverse and representative range of voices of those who are affected by and have an interest in the solution being developed and implemented (Turney 2014).

Co-production processes are particularly useful in the context of shaping responses to nexus shocks as they help to interpret this complex phenomenon whilst simultaneously considering and linking knowledge from local, regional, national and international levels and from actors working across energy, food, water and environment sectors (Corburn 2007). The approach is innovative in that it does not solely rely on scientific advice to inform decision-making and instead builds on the interactions and expertise of other audiences, acknowledging that the multitude of knowledge and expertise 'in the room' contributing to this process do so in a representative and effective manner. In so doing it builds stronger more trusted relationships between those taking part in the process who feel their contributions have been heard, that they are valued, and that multiple communities contribute to improving and implementing solutions (Ostrom 1996). This also enables flexibility and shifting of roles, for example, practitioners may go beyond their role of providers and recipients of knowledge to more active roles in designing evidence generation, ensuring a better alignment with end-user needs (Martin 2010).

The Nexus Shocks project adopted a co-production methodology which assumed that all those involved in the process had something to contribute even if not self-identifying as 'experts' (i.e. they still represented a type of 'non-expert' expertise based on their experiences of someone who would be affected by such a shock). This approach is based on the Nexus Network's definition of transdisciplinary research which 'not only integrates expertise from across academic disciplines, but also involves societal stakeholders in the design stage, and throughout the research process. … [since] knowledge can come from beyond formal academic disciplines, and insights are often provided through other kinds of tacit knowledge – as held by local communities, businesses, social movements or practitioners' (Cairns et al. 2017: 5). This process enables 'knowledge exchange and sharing of insights from a range of perspectives and expertise' (Howarth and Monasterolo 2017: 106). The co-production methodology was refined to specifically identify opportunities for improving responses to nexus challenges by innovatively assessing the complex nature of societal responses to these shocks to better inform business and policy responses that would be more favourable to bottom up solutions across the food-water-energy nexus.

To fully implement and benefit from a co-production approach, the Nexus Shocks project incorporated a rapid review of evidence on climate shocks, energy-food-water nexus, characteristics of resilience, and the nature of responses to these shocks. Following this, a small workshop was conducted with members of the project's Advisory Group to validate the research objectives, establish the programme of research, and refine the methodology (Howarth et al. 2015). Advisory Group members and workshop participants were selected from academia, policypractitioner, and industry communities, based on their knowledge, expertise and experience of climate change, the nexus, nexus shocks and responses to nexus shocks. Following this, five workshops (see Chapter 2) were conducted in London (UK) with a total of 78 stakeholders exploring a range of issues co-designed with workshop co-hosts: Predicting shocks and hazards (with the UK Met Office), Transmission and mitigation of nexus risks though infrastructure (with Atkins), Local economy responses to shocks (with Climate UK), Insurance and finance for resilience (with Lloyds of London and Willis Re), and Governance, governments and shocks (with Chatham House). Qualitative analysis was conducted of the findings from each workshop which were then discussed with workshop co-hosts and participants (through circulation and review of the first draft of the research report). Phase 2 of the project (the Nexus Shocks Fellowship) built on the findings from the workshops and conducted semi-structured interviews with key experts to further explore insights. Results were disseminated via a one-day symposium in London, project reports (Howarth 2016a, 2017a, b) academic papers (Howarth and Monasterolo 2016, 2017; Brooks and Howarth 2017), blogs (Howarth 2016d, 2017c) and policy briefs (Howarth 2016a, b, c; POSTnote 2016).

The co-production methodology was thereby used to design the research method for the project and became the subject of scrutiny and constructive dialogue between participants, enabling a strengthened process. Exploring the concept of co-production with Nexus Shocks participants saw a positive response to the methodology and active reflection on its validity and role in improving decision-making processes, collaboration and communication in response to nexus shocks. Co-production was however not necessarily understood in the same way by all those interviewed, with the majority considering it to be a loose way to capture whether or not end-users had been involved in evidence production. This differs from the discussion above which sees this as truly inclusive throughout the decision-making process, ensuring an iterative and reflective element and consistent communication and collaboration with all those involved. Its implementation however will vary depending on the nature of the project in question and stakeholders involved. A co-production process was described by a participant's experience as involving collaboration (and co-funding) between the Environment Agency, a Water Utility company and a city council working together on a flood alleviation project to collate evidence and knowledge on the physical environment, river flows and impacts on the community, subsequently making an integrated case for investment. Another example of a similar interaction entailed close collaboration and co-production of ideas between the Environmental Agency, National Trust, Natural England, and the Forestry Commission in response to a funding opportunity to support natural flood management. This process of engagement with multiple stakeholders, led by a funding body, ensured the close alignment of the funding available with community and stakeholder needs, to address important issues on flood management:

> (...) We are going to be, ultimately, some of the biggest end-users of a £4 million investment in natural flood management, and I think it was really helpful that [research council] approached us right from the very start. Before they'd actually got the programme sorted out, they were seeking our input in terms of planning, how it should work, what it should cover, what the call should set out, what priorities we had, right from the very start then. So, we had about a year's worth of planning before the call actually went out. So, we met with them two or three times and there's a workshop with academics that we attended as well. So, we were fully engaged in the process and I think that's been a really positive experience. (PO9)

For some, the way in which co-production is implemented was the standard way to develop a product or tool where an initial phase consists of the generation of ideas, followed by interaction with end-users to ensure the product meets their needs and is as helpful as possible. This would entail a robust process whereby insights at multiple stages are fed into facilitate and enable collaboration between experts and communities; create an advisory group for research council funded research programmes; develop responses based on local needs; or combine insights from different academic disciplines to understand how future trends will evolve (e.g. climate change and future security). This thereby enabled flexibility and agility and provided much clearer information to support responses. The inclusion of end-users is instrumental to co-production to ensure there is a close alignment between the (policy) decisions being made and the needs of those on the ground who will be affected. Whilst this was acknowledged and reflected in discussions with the various stakeholders through the duration of the Nexus Shocks project, there were mixed views about involving end-users as, whilst useful, it is not without its challenges.

Involving end-users enables access to valuable knowledge that would otherwise be untapped as it may be undocumented or anecdotal, for example, community knowledge on the location of vulnerable people and how to prioritize responses to heatwaves or flooding events. How end-users are defined also impacts how the co-production process will be designed. End-users could refer to community groups vulnerable to increasing threats of flooding and/or heatwaves in a given location, they could be local businesses, national networks, government departments formulating policies to increase resilience to climate risks, or indeed research funders seeking to improve the ways in which research they fund directly aligns with the needs of policy and business. In the latter, they act as a broker or intermediary end-user facilitating an extra mechanism in which to ensure alignment of research outputs on decision-making processes with broader, national mechanisms for national policy development. Co-production thereby enables closer collaboration between a range of actors, who may not necessarily have previously worked together. It can bring funding and funders together to develop innovative, agile and responsive modes of funding. For example, the UK's Economic and Social Research Council is committed to supporting research which has the potential for 'high scientific impact and/or high user impact' (ESRC 2017a: 15). One of the ways it achieves this is through funding recommendations being informed by a multidisciplinary grant assessment panel comprising both academics and a number of non-academic 'user' members from across the public, business and civil society sectors (ESRC 2017b).

A range of barriers to involving end-users also exist. Capacity, resources and misaligned timescales were the most frequently cited limitations, particularly when considering the complexity of shocks and the systems upon which they impact whose multiple stakeholders, cross-sectoral implications and conflicting values and priorities further act to exacerbate impact. Whilst short-termism dominates responses to climate shocks and can be perceived as a barrier, there are times where short-term and rapid responses are needed due to the dynamic nature of a shock and the rapidly evolving context of decision-making. Co-production is considered a constructive way to address complex problems where multiple sectors and stakeholders are involved but to be time and resource intensive. This is particularly the case in engaging and maintaining relationships between those involved and in sustaining the trust and willingness of end-users to engage. Whilst the process can lead to very positive experiences, there can be apathy to engage with a co-production process because it can be lengthy and benefits may not be immediately apparent. Individuals may not be able to recall their own experiences of the impacts of a climate event or how they responded or participated in imposed processes. Often cultural barriers emerge that need to be overcome across different stakeholder groups to ensure alignment of objectives and understanding of needs of those involved. This could entail not fully knowing who 'owns' the evidence available, who can access it, whose role it is to act upon it, or indeed how to act upon it. On this basis it can be practically challenging to identify and define who the end-users are.

CONCLUSION

Communication and collaboration between the public sector and civil society play an important role in the effectiveness of decision-making in response to nexus shocks through the provision of information, voluntary action and joint implementation of local projects. Communication on nexus shocks is challenged by the traditional linear dissemination process, and a lack of understanding of end-user needs. Improved communication on nexus shocks can be achieved through the contextualization of warnings, providing digestible, targeted, clear and consistent messaging on the expected impact and required response, using language which is relatable, and ensuring effective dissemination structures are in place. On the other hand, collaboration can increase the effectiveness of responses by providing a forum for constructive conversations, facilitating the exchange of information and better access to data, evidence and expertise, and stimulating better frameworks to align decision-makers and end-user needs. Good collaboration takes time and depends on trusted relationships, where collaborators see each other as partners rather than stakeholders, adopting joined-up, flexible and agile processes, aligned with the needs of those affected by nexus shocks.

In particular, co-production provides a constructive way to address complex problems, which incorporates cross-sector and stakeholder priorities and processes. This ultimately leads to better alignment with the needs of end-users, resulting in more salient decision-making processes for those implementing resilience to nexus shocks on the ground. However, decisions made to shape responses to nexus shocks formulated at the national level may fail to consider implications for the local level, longer term, or broader sector implications. The examination of capacities across scales can help identify commonalities leading to a clearer and more relevant framing of transformative social change. Nevertheless, there is often a misalignment between the needs of decision-makers at the local level and evidence and scientific information available to help inform their decision-making processes (which are channelled through national processes) which must subsequently be addressed.

REFERENCES

ASC. (2016). *UK climate change risk assessment 2017 synthesis report: Priorities for the next five years*. London: Adaptation Sub-Committee of the Committee on Climate Change.

Bammer, G. (2013). *Disciplining interdisciplinarity: Integration and implementation sciences for researching complex real world problems* (472 pp.). Canberra: ANU Press.

Bennett, J. B., Blangiardo, M., Fecht, D., Elliott, P., & Ezzati, M. (2014). Vulnerability to the mortality effects of warm temperature in the districts of England and Wales. *Nature Climate Change, 4*, 269–273.

Bidwell, D., Dietz, T., & Scavia, D. (2013). Fostering knowledge networks for climate adaptation. *Nature Climate Change, 3*, 610–611.

Brooks, K., & Howarth, C. (2017). Decision-making and building resilience to nexus shocks locally: Exploring flooding and heatwaves in the UK. *Sustainability, 9*, 838.

Cairns, R., Wilsdon, J., & O'Donovan, C. (2017). *Sustainability in turbulent times: Lessons from the Nexus Network for supporting transdisciplinary research*. Brighton: The Nexus Network.

Christel, I., Hemment, D., Bojovica, D., Cucchiettia, F., Calvo, L., Stefaner, M., et al. (2017). Introducing design in the development of effective climate services. *Climate Services, 9*, 111–121.

Clarke, B., Stocker, L., Coffey, B., Leith, P., Harvey, N., Baldwin, C., et al. (2013). Enhancing the knowledge governance interface: Coasts, climate and collaboration. *Ocean and Coastal Management, 86,* 88–99.

Corburn, J. (2007). Community knowledge in environmental health science: Co-producing policy expertise. *Environmental Science & Policy, 10*(2), 150–161.

ESRC. (2017a). *ESRC research funding guide October 2017.* Economic & Social Research Council. Available online at http://www.esrc.ac.uk/files/funding/guidance-for-applicants/research-funding-guide/.

ESRC. (2017b). *ESRC responsive mode grant assessment process.* Economic & Social Research Council. Available online at http://www.esrc.ac.uk/about-us/governance-and-structure/advisory-committees/research-committee/responsive-mode-grant-assessment-process/.

Hine, D. W., Reser, J. P., Morrison, M., Phillips, W. J., Nunn, P., & Cooksey, R. (2014). Audience segmentation and climate change communication: Conceptual and methodological consideration. *WIRES Climate Change.* https://doi.org/10.1002/wcc.279.

Howarth, C. (2016a). *What we've learnt so far: Findings from the nexus shocks network.* Global Sustainability Institute report. UK: The Nexus Network. Available online at http://www.thenexusnetwork.org/wpcontent/uploads/2016/01/Nexus-Shocks-Network-What-We-Are-Learning_CandiceH.pdf.

Howarth, C. (2016b). *Responding to extreme weather events.* ESRC Evidence Briefing. Swindon, UK: ESRC. Available online at https://esrc.ukri.org/news-events-and-publications/evidence-briefings/responding-to-extremeweather-events/.

Howarth, C. (2016c). *Informing decision-making in response to nexus shocks* (LWEC PP Note).

Howarth, C. (2016d). *The nexus shocks network.* Nexus Network. Available online at www.nexusnetwork.org.

Howarth, C. (2017a). *Informing societal responses to shocks to the energy-food-water nexus: The nexus shocks network* (Report to the ESRC Nexus Network).

Howarth, C. (2017b). *Nexus Network fellowship: Nexus shocks summary of findings, 2017.* University of Surrey/The Nexus Network. Available online at http://www.thenexusnetwork.org/wp-content/uploads/2018/02/Howarthreflective-report-to-Nexus-Network-Sept-2017.pdf.

Howarth, C. (2017c). *The nexus shocks fellow.* Nexus Network. Available online at www.nexusnetwork.org.

Howarth, C., & Monasterolo, I. (2016). Understanding barriers to decision-making in the UK energy-food-water nexus: The added value of interdisciplinary approaches. *Environmental Science & Policy, 61,* 53–60.

Howarth, C., & Monasterolo, I. (2017). Opportunities for knowledge co-production across the energy-food-water nexus: Making interdisciplinary approaches work for better climate decision-making. *Environmental Science & Policy, 75,* 103–110.

Howarth, C., & Painter, J. (2016). The IPCC and local decision-making on climate change: A robust science-policy interface? *Palgrave Communications, 2,* 16058.

Howarth, C., Jones, A., Philip, G., & Hogbin, J.-A. (2015). *Nexus Shocks network: Decision-making on nexus shocks.* Summaries of workshop discussions, Global Sustainability Institute, Anglia Ruskin University.

Howarth, C., Viner, D., Dessai, S., Rapley, C., & Jones, A. (2017). Co-producing climate change knowledge: Incorporating practitioner evidence in the IPCC WGII process. *Climate Services, 5,* 3–10.

Ingold, K., & Fischer, M. (2014). Drivers of collaboration to mitigate climate change: An illustration of Swiss climate policy over 15 years. *Global Environmental Change, 24,* 88–98.

Jasanoff, S. (2010). A new climate for society. *Theory, Culture & Society, 27*(2–3), 233–253.

LCCP. (2018a). *About.* Accessed 22 February 2018. Available online at: http://climatelondon.org/lccp/.

LCCP. (2018b). *Heatwaves.* Accessed 22 February 2018. Available online at: http://climatelondon.org/climate-change/heatwaves/.

Lemos, M. C., Kirchhoff, C. J., & Ramprasad, V. (2012). Narrowing the climate information usability gap. *Nature Climate Change, 2,* 789–794.

Löhr, K., Weinhardt, M., Graef, F., & Sieber, S. (2017). Enhancing communication and collaboration in collaborative projects through conflict prevention and management systems. *Organizational Dynamics.* In press. https://doi.org/10.1016/j.orgdyn.2018.05.002.

Martin, S. (2010). Co-production of social research: Strategies for engaged scholarship. *Public Money & Management, 30*(4), 211–218.

Met Office. (2015). *The heatwave of 2003.* Accessed 22 February 2018. Available online at: https://www.metoffice.gov.uk/learning/learn-about-the-weather/weather-phenomena/case-studies/heatwave.

Nerlich, B., Koteyko, N., & Brown, B. (2010). Theory and language of climate change communication. *WIRES Climate Change, 1,* 97–110.

Newton, J. (2017). Shared intentions: The evolution of collaboration. *Games and Economic Behavior, 104,* 517–534.

Omondi, F. O., Hayombe, P. O., & Agong, S. G. (2014). Participatory and innovative design guidelines to planning and managing urban green spaces to transform ecotourism. *International Journal of Current Reasearch, 6*(12), 10397–10402.

Ostrom, E. (1996). Crossing the great divide: Co-production, synergy and development. *World Development, 24*(6), 1073–1088.

Pidgeon, N., & Fischhoff, B. (2011). The role of social and decision sciences in communicating uncertain climate risks. *Nature Climate Change, 1,* 35–41.

Polk, L. (2015). Transdisciplinary coproduction: Designing and testing a transdisciplinary research framework for societal problem solving. *Futures, 65,* 110–122.

POSTnote. (2016, December). *The water-energy-food nexus* (POSTNote 543).

Scandelius, C., & Cohen, G. (2016). Sustainability program brands: Platforms for collaboration and co-creation. *Industrial Marketing Management, 57,* 166–176.

Tàbara, D. J., St Clair, A. L., & Hermansen, E. A. T. (2017). Transforming communication and knowledge production processes to address high-end climate change. *Environmental Science & Policy, 70,* 31–37.

Turney, J. (2014). *Future earth blog.* Last accessed 13 October 2015. Available online at: http://www.futureearth.org/blog/2014-jul-23/be-inclusive-you-need-more-voicesqa-sheila-jasanoff.

Vesselinov, X., & Zhang, V. V. (2016). Energy-water nexus: Balancing tradeoffs between to level decision-makers. *Applied Energy, 183,* 77–87.

Voorberg, W. H., Bekkers, V. J. J. M., & Tummers, L. G. (2014). A systematic review of cocreation and co-production: Embarking on the social innovation journey. *Public Management Review, 17*(9), 1333–1357. https://doi.org/10.1080/14719037.2014.930505.

Watson, V. (2013). Co-production and collaboration in planning—The difference. *Planning Theory & Practice, 15*(1), 62–76.

Webber, S. (2017). Circulating climate services: Commercializing science for climate change adaptation in Pacific Islands. *Geoforum, 85,* 82–91.

CHAPTER 5

Looking Ahead

Abstract The frequency and intensity of climate shocks are expected to increase under a changing climate with severe implications for sectors and those working across the food, energy, water, environment nexus. Impacts of these shocks will exacerbate the vulnerability of those sectors affecting resource availability, system pressures and decision-making processes. We reflect here on how communication, collaboration and co-production can play a fundamental role in informing nexus related decision-making and increasing resilience to shocks and discuss how mechanisms through which stakeholders working, across the nexus (e.g. on energy, food, water, environment) can more efficiently and more robustly co-create robust responses to nexus shocks. Fundamental to embedding communication, collaboration and co-production within responses to nexus shocks and building resilience is the availability and deployment of sufficient financial resources and capacity building in order to facilitate this process and ensure it is sustained in the long term.

Keywords Embedding practice · Co-production · End-users Climate change · Resilience

© The Author(s) 2019
C. Howarth, *Resilience to Climate Change*,
https://doi.org/10.1007/978-3-319-94691-7_5

HIGHLIGHTS

- Climate shocks will become more frequent and intense under a changing climate with significant implications for the food, energy, water, environment nexus, calling for communication, collaboration and co-production to be embedded within decision-making and resilience building.
- Improving resilience to climate shocks, and their impacts on the food, energy, water, environment nexus should focus on responses, governance, communication, collaboration, co-production and support.
- Embedding co-production in research development, responses to nexus shocks and building resilience to climate shocks requires adequate funding and capacity building in the long term.

BUILDING RESILIENCE TO NEXUS SHOCKS

Sound decision-making processes require structures, communication channels, actors and policies to be resilient and operational during a nexus shock event and following its occurrence. In addition, contingency measures are required to be put in place to account for any loss of historical or institutional memory or gaps in expertise, knowledge, training and skills that could otherwise be invaluable during shock events (Howarth and Monasterolo 2016). With this in mind, and considering the complexities of impacts and responses to nexus shocks seeing the variety of stakeholders involved, Cash and colleagues (2012) raise important questions about how to assess the extent to which a decision-making process is fair and legitimate with experts involved throughout, and where fairness could be considered as elementary to 'good' decision-making which then raises the question of how 'good' decision-making is judged and by whom. Through decision-making processes, impacts can become more apparent and roles evolve to reflect key resource and expertise needs. With specific reference to the context of nexus interactions, the term resilience can have different meanings for different stakeholders (Wentworth 2016). However the ability of a system to withstand external shocks is critical as this determines its vulnerability, whilst bearing in mind the implications of being resilient to one shock which can result in other system's or stakeholder's resilience being affected. This is an important point to consider in assessments of decisions made regarding nexus shocks to ensure a holistic approach whereby all interests are considered in developing resilient responses. The concept of resilience emerges primarily from a context of disasters, and as a result of this, resilience has been framed as a four-stage process (DFID 2011):

- *Context*: what is the focus of resilience and whose resilience is being built or affected (depending on geographic location, social groups involved, socio-economic systems, environmental context or institution)?
- *Disturbance*: what is the type of shock that a system needs to build resilience to?
- *Capacity to deal with disturbance*: what is the ability of a system or process to cope with a shock or stress? This depends on exposure (magnitude of the shock), sensitivity (the degree to which a system will be affected by or will respond to a given shock) and adaptive capacity?
- *Reaction to disturbance*: what are the different possible outcomes from survival mode to transformation mode?

When exploring resilience of systems, in the case of nexus shocks in this book, there are important questions to consider: Resilience *to what* (e.g. What types of threat, shock or stress)?; Resilience *of what* (e.g. a structure of function; a community, an institutional structure or infrastructure or the qualities of services being resilient? and Resilience *for whom* (i.e. who is likely to benefit or who may be excluded)? As discussed in this book, current UK government policy mainly adopts a reactive decision-making process to build resilience *to* specific climate or severe weather-related shocks and on *resilience of* a wide range of entities such as communities, institutions and structures (Twigger-Ross et al. 2015; Howarth and Monasterolo 2016; Johnson and McGuinness 2016). However, events such as these that often require an emergency response, are not necessarily assessed and responded to pro-actively. There is a need to adapt and transform to future change in economies, population, governance and climate shocks with implications for the food, energy, water, environment nexus in order to fully incorporate the importance of action and responses at multiple governance levels (Howarth and Monasterolo 2016). However there are questions about the ability of governmental structures to plan in advance to effectively respond to such emergencies, extreme events or shocks and to support community and system resilience if events are uncertain and unpredictable (Pelling and Dill 2010). When exploring this in the context of complex adaptive systems (CAS), Duit and Gallaz (2008) identify five key characteristics for effective governance structures and resilience: (i) *diversity* of actors and structures in governance; (ii) *autonomy of* actors and structures; (iii) *interdependence* of actors and structures; (iv) *adaptability* of actors and structures to learn from experience and each other; and (v) *collaboration* between actors and institutions. This further highlights the importance and role of strong collaborative processes and connections between cross-sectorial actors and institutions for better, more informed decision-making and building of resilience in response to a nexus shock.

Climate shocks are predicted to increase in future due to a changing climate, increased exposure, growing population and greater impacts on vulnerable groups, which calls for more effective, joined-up, integrated decision-making across the nexus sectors and stakeholders. The frequency and intensity of nexus shocks are predicted to increase both globally and in the UK (ASC 2016), and, being characterized as unpredictable and uncertain with a web of complex impacts, effective responses will require communication and collaboration across different levels of society, governments and the scientific community. We have seen that decision-making will be at its most effective in this context if it is a process which involves multiple people, organizations, sectors and strategies, creating opportunities to examine issues within these decision-making processes that occur during nexus shocks and enabling resilience to shocks.

Heatwaves are currently rare in the UK, however, it is predicted that the frequency, intensity and length of heatwaves will be made worst by future changes in climate demographics and urban development (ASC 2016; Smith et al. 2016; Burchell et al. 2017). Public Health England's (PHE) UK Heatwave Plan has been published annually since 2004, following the heatwave experienced across Europe in August 2003. The shock of a severe and prolonged heatwave such as this, as explored in this book, can negatively impact food, water and energy resources as well as businesses, transport, health and social care services with effects felt at the local level across this nexus. The 2017 UK Climate Change Risk Assessment expects that heatwaves in the summer, similar to the one experienced in 2003, are likely to become the norm by the

2040s (ASC 2016). People, systems and structures are particularly vulnerable to high temperatures, due to increased exposure and sensitivity to high temperatures, exacerbated by the quality of housing and the built environment, local urban geography, household income, employment, tenure, social networks and self-perception of risk (Benzie et al. 2011; Benzie 2014). Increased exposure is also closely associated with increased outdoor activity during periods of hot weather with a lack of individual action to protect against the heat (Lefevre et al. 2015) as heat risks tend not to be perceived as personal risks (Wolf et al. 2007).

As discussed throughout this book, we know that food, energy, water and environment resources and systems are deeply linked. The concept of 'nexus', which is widely used, can provide a useful lens through which to explore the interlinkages and interdependencies between these resources however the meanings, applications and implications of using the term 'nexus' are contested and varied in this context. This book has explored nexus *shocks* which we have defined as any sudden occurrence that disrupts and has knock-on effects across the food–energy–water–environment nexus and/or the actors that work within it. These can vary in nature and for the purpose of this book, we have focused specifically on extreme weather events or shocks resulting from climate variability and change. When considering the complexities and intricacies of the nature of nexus shocks, their impacts and responses across the food, energy, water, environment nexus, there is a need for an integrated approach at multiple levels across sectors and systems in preparing, planning, identifying, responding and making the most appropriate decisions to nexus shocks that incorporate the interlinkages between these core resources. However, a national strategy to build resilience to nexus shocks is missing, meaning that climate vulnerable places, people and the systems upon which they depend may not be adequately targeted and supported. Responses to climate shocks can exacerbate initial impacts across the energy, food, water, environment sectors if these are not adequately managed, however, these factors can also help mitigate future shocks. Factors that make the impact of shocks worst (e.g. exacerbating factors), when properly understood, identified and implications examined, may not be fully or permanently removed, and can become obstacles that exist within the system. The impacts of climate shocks to the food, energy, water and environment nexus demonstrate that there are similarities when explored through an exacerbation/mitigation lens, in terms of the prediction of shocks, interactions with infrastructure, local shock experiences, the role of finance and insurance, and governments and governance.

A range of challenges emerge from nexus shocks, including the interconnected nature of systems which are impacted, communication and collaboration processes in place between stakeholders, the nature of decision-making processes, existence of weak links between sectors, stakeholders and components within a system, over-reliance on resources which are vulnerable to these shocks, social and cultural factors and the nature of responses to nexus shocks. However, in light of this, a range of opportunities emerge from nexus shocks which can materialize from these challenges: strategic thinking takes into consideration any limitations experienced, collaboration and communication can be improved, societal responses to these shocks can be better anticipated, processes are improved, the vulnerabilities of key systems can be better understood and identified, and ultimately, capacity to respond and build resilience can be built up. There are strong similarities and overlaps in the challenges and opportunities identified in the context of climate shocks to the food, energy, water, environment nexus, a thorough understanding is therefore needed of the relationship between society, the system on which it depends and its components (e.g. infrastructure, healthcare etc.), and how these are affected by the impacts from shocks. This will ultimately enable a better assessment of how a nexus shock can affect resilience and capacity to respond. When considering individual nexus shocks such as heatwaves and flooding, increased resilience to these shocks requires effective communication and collaboration. Better communication can be achieved by contextualizing warnings, providing digestible, targeted, clear and consistent messaging, using relatable language, and ensuring effective dissemination structures are in place. Similarly, good and effective collaboration requires additional time to focus on responses and roles, trusted relationships, collaborators as partners, and joined-up, agile processes, aligned with the needs of those affected. Co-production provides a constructive way to deliver this and provide more salient decision-making processes which better incorporate the needs of different stakeholders and actors who are affected by nexus shocks and whose decisions will determine how resilient responses are.

We have seen how factors that exacerbate the damaging impacts of nexus shocks (such as timing, collaboration, evidence, culture, responsibility and responses, and technology) can also play a role in mitigating them. Limitations in science and the ability to adequately predict the timing, duration, location and impacts of shocks can have negative implications for the credibility of decision-making processes further compounded by limited time and resource to respond or an abundance of time and resource committed when no shock materializes. The involvement of numerous actors is required to ensure collaboration best represents the needs and interests of those affected by these shocks. However the involvement of multiple individuals and organizations can lead to competing demands with potential impacts on increased likelihood of cascading shocks across other sectors ultimately increasing the vulnerability of actors, resources, processes or infrastructure within the nexus.

Nexus shocks are characterized by how they affect food, energy, water, environment resources as well as the web of interactions between stakeholders, sectors and decision-making processes within them. This implies a mix of cultures, behaviours, values and priorities which can further affect the responses that occur and how external impacts can exacerbate them, particularly when nexus shocks are experienced differently by different sectors and engender different, sometimes conflicting, responses. Whilst identified as a barrier to decision-making, collaboration and communication are also seen as an opportunity to strengthen relationships, knowledge exchange and decision-making processes. Communication and collaboration can be seen as a barrier, however if their sources, nature and impacts are better understood then common languages and a joint vision between stakeholders can be developed to establish mechanisms through which to communicate effectively and resources pooled to target where needed. This vision can ensure better strategizing to filter and prioritize evidence to inform decision-making, co-produce evidence where needed and target this to specific audiences based on the context, impacts experienced, needs and resources available.

Adopting an iterative and reflective process, whereby systems in place and the stakeholders working within them actively seek to self-evaluate their progress and work and embed learnings into their decision-making processes, can ensure a smoother flow of information to underpin effective communication and explore the range of scenarios of impacts, responses and vulnerabilities that could ensue from a particular shocks. Feeding this into decision-making processes can increase the credibility of responses initiated, facilitate and better manage the flow of evidence produced and used, improve trusted relationships between stakeholders and limit costs and risks of the initial shock to a system. The utilization of new and innovative methods can be an effective way to do so in order to ensure timely and instant delivery of evidence whilst accounting for potential risks of mismanagement or overuse. This can thereby contribute to building a powerful narrative on how nexus shocks impact food, energy, water, environment resources and stakeholders, demonstrating the benefits of collaboration, communication and knowledge exchange.

BUILDING COLLABORATIVE PROCESSES

A nexus approach needs to provide innovative approaches and insights on the impacts of extreme weather and climate-related shocks, such as heatwaves and flooding, to energy, food, water, environment resources. Doing so in a collaborative way which enables knowledge exchange from across disciplines (such as climate science, economics, social science, engineering), can help encourage effective decisions and responses by sharing knowledge, skills, expertise, best practices and lessons learned. The widespread assumption that providing more information will lead to desired change is flawed. Instead, an emphasis on devising and implementing solutions illustrates the need for mechanisms to facilitate decision-making processes. Climate services for example, which are co-developed with end-users (Christel et al. 2017), are useful ways of delivering information directly aimed at apprising decision-makers (Webber 2017), provide appropriate representation of related governance processes (Clarke et al. 2013), and address potential issues of malpractice (Scandelius and Cohen 2016). In doing so, collaborative processes, built on humans' natural ability to collaborate for mutual gain (Newton 2017), actively pursue mechanisms for conflict resolution (Löhr et al. 2017) and build on and acknowledge the need for more 'complex forms of agent interactions in the production, framing, communication and use of climate knowledge. In particular, these provide specific procedures able to tackle difficult normative questions regarding assessment of solutions and the allocation of individual and collective responsibilities' (Tabara et al. 2017: 31). Beliefs among stakeholders and formal power structures (as opposed to structures of power that are perceived) play an important role in decision-making processes for responses to climate change (such as climate mitigation) whilst proving limited in the implementation of these responses (Ingold and Fischer 2014).

A shock provides an opportunity to bring stakeholders together to target specific responses in the immediate aftermath of the shock, and to plan pro-active responses for future shocks. More active collaboration between policymakers, practitioners and academics as partners rather than stakeholders can facilitate more joined up approaches with greater ability to influence. This can be further strengthened by adopting flexible and agile processes where contributors benefit from innovative ways of working and sharing in a way that aligns with the needs of affected communities. This enables a stronger focus on working backwards from the desired outcome, taking pro-active and pre-emptive approaches for shock events, with the bigger and longer-term picture in mind. Collaboration between nexus stakeholders enables a more comprehensive view of the shock at hand, captures the needs and objectives of stakeholders affected by shocks and leading responses, and enables a rapid assessment of potential impacts to infrastructure, human lives, economy, culture and the environment. In doing so, it facilitates the exchange of information and better access to data, evidence or expertise providing a rich picture of what is happening on the ground, enabling a hands-on view of how a system works and brings together practical expertise and theoretical knowledge, enhancing the quality of data, relationships and responses.

EMBEDDING COMMUNICATION OF CLIMATE CHANGE IN RESILIENCE

The information deficit model, discussed in this book, assumes an audience does not have sufficient knowledge about an issue (e.g. scientific and technological problems) and hence that (scientific) experts should produce that expertise, on the assumption it is correct and share what they consider is needed for lay audiences to fully understand the issue. This assumes scientific input needs to be developed independently and fed into political decision processes as opposed to being informed by it. However, the interaction between knowledge provision and decision-making and behaviour change is complex and begins with the assumption that scientific knowledge is the most useful (and often perceived as the only) source of information. In fact, in an assessment of the value of (public) engagement with science, Stilgoe and Wilsdon (2014), claim that 'public engagement is a necessary but insufficient part of opening up science and governance' (p. 4) and that historically there has been an over-emphasis on the means of engagement rather than the aim. This relies on assumptions from the information deficit model which aims to reach a consensus which 'can disguise the diversity of views that are likely to define a particular issue' (p. 6). However, this is a necessary process agreed upon by scientists, funders, policy makers yet 'the rapid move from doing communication to doing dialogue has obscured an unfinished conversation about the broader meaning of its activity' (p. 8). An important limitation here is the use of three frames in the information dissemination approach ('if only they knew', 'if only they could be made to care', 'if only they could stay home') when in fact framing the issue should be around the 'social potential' that exists which enables opportunities for action beyond individual choices that people make and 'to the creative abilities of people to participate in change in ways that fits their own contexts and concerns' (Moezzi and Janda 2014: 31).

Important issues around communication of climate change, in order to underpin responses to climate shocks with up to date evidence, rest with an ever-evolving science and corresponding economic, social, cultural and political contexts. This leads to a complicated language used to transfer information. Consistent interaction, dialogue and collaboration between scientists and policy makers is therefore required to ensure decision-making (Bidwell et al. 2013) reflects the most up to date scientific evidence. However, there are numerous issues with this: building strong trusted relationships takes time as well as skills and an appetite for relationship-building which may be lacking; decision-makers will need to consider numerous other factors in their decisions in addition to climate, and the dialogue is not linear containing many complexities. Individual's decisions rest on a web of values, beliefs, preconceptions and more, consequently, different strategies are required to address these various values, beliefs, motivations, perceptions, attitudes and stakes involved. Communication processes should also be based on a sound understanding of social learning, with 'the movement of information and practices through knowledge networks, whose structures influence both the pace and qualities of learning as the networks themselves evolve' (Bidwell et al. 2013: 610). Nowadays, information on climate change is better known, understood and more relevant to different audiences, across different media (e.g. press, film, literature, theatre, policy etc.) yet emissions continue to rise, impacts of climate shocks on the food, energy, water, environment nexus are more prominent and resilience to these shocks remains in a state of progress. This leads to a questioning of the efficacy of what is communicated, the messenger and the ability of the audience to translate their acquired awareness into action.

Whilst, communication and collaboration are key to ensure resilience to nexus shocks (Howarth and Monasterolo 2016) this will be most effective when combined with a deeper understanding of the barriers to change that exist. This will ensure decision-making processes align better with the needs of end-users and those working across the nexus, without the risk of neglecting certain impacts or stakeholders. With this in mind, the use of language is particularly important in communication and framing, as terms like 'climate change' and 'nexus' are difficult for audiences to connect to, however talking in terms of 'hot weather' is immediately relatable. The challenge posed by the range of languages and lexicons that are used in different organizations to describe similar concepts, can demonstrate credibility and trust, however, this can lead to an immobility of action. Proper structures must, therefore, be in place to share and disseminate effectively through a range of communication channels and media available, ensuring accurate and clear information is feeding through planning, preparation, and response stages of a shock.

Communication and collaboration play an important role in responding to nexus shocks and can be both a barrier and an opportunity to ensure resilience to shocks in this regard. This is especially important when considering the characteristics of the nexus and shocks that impact it with divergent and, at times, conflicting impacts, interests, stakeholder and societal needs and decision-making processes. A consistent, transparent and reflective flow of information is needed to ensure parties involved acquire the most up to date knowledge on impacts and responses in order to formulate effective decision-making processes and ensure a sustainable and effective interactive process between scientists, practitioners and policy makers. This further ensures that certain barriers such as lack of public understanding of risks, desensitization, and reduced trust (Howarth and Monasterolo 2016) are brought to the fore and addressed from the outset to encourage more inclusive, pro-active and collaborative processes with a system-wide view are in place. Doing so therefore tackles these challenges directly and reframes them as opportunities to improve decision-making, responses and efforts to increase resilience to nexus shocks.

Decision-makers require a better understanding of the contextual factors that affect how a shock impacts the nexus and its stakeholders, strategic thinking requirements to co-develop visions and processes for decision-making, the range of challenges and opportunities that emerge from collaboration and communication during and in the aftermath of a shock, and timings and processes for decision-making across sectors that span the nexus. Due to the complexity and cross-boundary nature of nexus shocks, national and local policy responses to extreme weather and climate events should apply a 'whole systems' approach, taking into consideration how decisions are influenced by and affect various stakeholders and sectors. Within this, there is a need for greater policy emphasis on strategic and proactive measures to mitigate extreme events, rather than relying on reactive policies after the event. Policymakers should, therefore, engage in information-sharing and knowledge integration with other countries and regions experiencing similar shocks to learn from their experiences and share best practice. Through this, a better re-negotiation is needed of existing 'social contracts' between society, assets and infrastructure, and with stakeholders involved in the nexus—enabling ordinary civilians to play a more active role in shaping responses to nexus shocks and reflecting on their own practices, dependencies and relationships with these assets. By adopting such a range of integrated strategies needed to engage and adopt cross-stakeholder learning will enable such reflective approaches to better design and implement processes in response to nexus shocks. This could take the form of a national task force focusing on co-production of approaches, secondment schemes and stakeholder engagement at the local, national and international scales, fully engaging with the key agencies and stakeholders that are affected by and responding to extreme weather events (Howarth 2016).

Co-Production and Engaging End-Users

Co-production approaches are referred to as those that are flexible enough to adapt to the evolving nature of challenges society faces, which requires collaborative processes that are relevant and can lead to the (co-) production of real world solutions and fill known evidence gaps. Co-production enables both inclusive and self-reflective approaches to merge to enable a closer alignment between decision made and those implemented or affected by these decisions. This occurs whilst working with the set of challenges the process faces whilst acknowledging the opportunities this provides. This is an innovative approach which draws upon a number of academic disciplines enabling stakeholders, organizations and systems working in siloes to participate in a process where interrelationships and complexities within a system (such as the food, energy, water, environment nexus) can be better understood (Bammer 2013). Co-production processes are not entirely new and have materialized in areas of climate change, ecosystems services, public policy, community engagement and environmental challenges. It has gained prominence due to a focus on end-users (Voorberg et al. 2014), and incorporating actors beyond the typical players academia (Polk 2015), which enables closer alignment and relevance to the rapidly evolving policy landscape, resource needs and capacity restrictions that characterize response to climate shocks to the nexus. As a process, it is characterized by being inclusive, collaborative, integrative, usable and reflective (Polk 2015) making it a fundamental process to democratize politics (Jasanoff 2010) and incorporate a broader set of diverse voices. Ultimately, this enables the co-creation of both a solution and a process to address a particular challenge (Turney 2014).

Co-production provides space for knowledge exchange and sharing of insights from a range of perspectives and expertise. Fundamental to this process is the acknowledgment that of all those who contribute to the process no one and everyone is an expert in something (Howarth and Monasterolo 2017). The process helps to interpret complex phenomena whilst linking local, multi-sector knowledge to national and global social change movements, with a variety of components allowing for the incorporation of knowledge from across scales (Corburn 2007) making it an innovative and fitting approach to consider complex nexus related issues. Durose et al. (2011) state the importance of including non-experts in deliberative processes on issues of relevance to different communities by empowering them in the process (Collins and Evans 2002), and to include arguments that may have been overseen by experts (Fisher 2000) and ultimately enabling these communities to contribute to improving outcomes and implementing solutions (Ostrom 1996). Involvement of practitioners in co-production processes thereby forces them to go beyond their role as providers and (passive) recipients of knowledge to those who play an active role in commissioning, overseeing and assessing evidence. This can lead to higher levels of engaged and utilized work and better alignment with end-user needs (Martin 2010).

Peer to peer dialogue enables better sharing of evidence, responses and information on the culture, values and priorities of those part of the process informing decisions to improve resilience to nexus shocks. Consideration of the context of decision-making, particularly the current and future risks, impacts experienced, end-user needs and responses strengthens these peer to peer interactions enabling individuals, organizations and these processes to become conduits for solutions. In addition, this can enable a stronger mechanism through which up to date information about projects and processes under development can be shared and thereby enable a faster deployment of solutions and responses where needed. Engagement and collaboration with end-users from the outset is instrumental to ensure solutions and response to climate shocks are robust, effective, and reflect the needs, abilities and vulnerabilities of those impacted by shocks. Work in this manner, where collaboration enables the building and strengthening of *partnerships*, is central to a number of frameworks to develop guidance on responses to shock scenarios, indeed this enables immediate responses to take place when no alternative exists providing a platform from which co-production of resilience becomes a joint effort. This, therefore, ensures a better alignment with these partners and relevant stakeholders when a shock occurs, building trusted relationships that facilitate future decision-making. Whilst this can require a long drawn-out process in order to build trust across stakeholders, this is overly beneficial in the long run as relationships are already in place when they need to be drawn upon and in the event of staff turnover, any loss of expertise or link with a particular partner can be more readily re-established.

End-user engagement should start from seeking to understand current concerns and perceptions of barriers to change (Cruickshank 2017) and be conducted in a flexible, reflective and iterative manner enabling stakeholders to choose when to participate in partnership processes when needed and convenient to them. Removing a sense of exclusivity enables freedom to ensure knowledge exchange on existing and upcoming shocks and related developments which guarantees more sustained and committed engagement. The EU-funded project 'RESIN' provides an innovative example of this (EU 2016): consisting of 17 partners from eight European countries and consisting of knowledge institutes, universities, consultancies (ARCADIS), cities (Manchester, UK; Bratislava, Slovakia; Paris, France; Bilbao, Spain), a network organization (ICLEI) and a standardization institute enables the project to develop innovative methodologies and decision support tools for cities which are likely to experience more extreme weather events due to climate change. The project aims to co-create these new methodologies and tools by facilitating a close working relationship between the research and practice partners to ensure these are tailored to the needs of each end-user (cities). At its centre is a process of and capacity building through workshops with city practitioners, local government representatives, infrastructure providers, researchers and climate adaption experts to enable knowledge exchange and the building of tools for future co-produced climate adaptation solutions in urban contexts. It presents, therefore, an effective model going beyond peer-to-peer approaches, and through which systems can learn from other systems, for example where learning arises across stakeholders within cities and across cities whose vulnerabilities and responses to nexus shocks may vary but provide valuable lessons to others.

Whilst this should be embedded in response to shocks, the impact and evaluation of these interactions should also be an important component. Evaluating the impact of end-user engagement can be challenging to quantify as engagement in such a process may not lead to tangible outcomes in the short-term whilst still building longer-term effective collaborative processes which may be drawn upon on indirect topics and for future needs. Whilst this may not be quantifiably measurable, the benefits of knowledge exchange and engagement with new and different stakeholders facilitates a process by which each is aware of the current and future needs of end-users within the process, and particular responses or shocks may change in relevance to specific end-users depending on the context.

IMPROVING NEXUS SHOCK DECISION-MAKING-PROCESSES

Due to the collaborative nature of managing and responding to nexus shocks, a significant amount of information will be shared with partner organizations, agencies or bodies. However, the extent to which evidence is useful is affected by issues around accessibility and data sharing, a lack of technical capacity to utilize existing data, a disconnect between evidence producers and end-users, and difficulties in operationalising evidence in decision-making. This can lead to hindrances to decision-making due to the levels of accuracy or bias in evidence available, problems with conflicting information, timing or the lack of dynamism, and evidence misused or not being used. There are times when judgement-based decisions are inevitable and can be beneficial, however, (internal) politics can often be a driving factor, at times related to a reluctance to act or to apply lessons. This conflicts somewhat with the perceived benefits of evidence-based approaches being the lack of ambiguity, shared agreed decisions, clear expectations regarding processes, the removal of argument and debate, which can lead to more streamlined, replicable, efficient decision-making processes. The production and use of evidence is underpinned by clear and effective communication approaches to specific audiences which can ensure progress in resilience is supported by adequate and relevant information. Generally the communication of evidence from experts and scientists is considered good and joined up, particularly for operational responses as opposed to preparation and resilience. However, information available to the public and decision-makers is considered less effective, particularly when communicating risk which tends to incorporate a mix of data on frequencies, probabilities, return periods and statistics which do not resonate with all audiences. Different risks need to be communicated in different ways and are received differently by audiences: heat risks for example pose a challenge as they are inherently difficult to communicate and audiences often enjoy the heat and sunshine as there are no apparent immediate physical risks.

The Nexus Shocks project provided important insights which guide a set of recommendations for decision-making in the context of climate shocks and increasing resilience across the food, energy, water, environment nexus (Table 5.1). This highlighted, in particular, the importance of collaboration. Whilst collaboration between stakeholders occurs during nexus shocks, this is often reactionary, and building resilience would be further improved if collaboration were embedded in preparatory processes in between shocks. In so doing, this would enable a comprehensive picture of a shock or response to be constructed and coordinated enabling a consistency of approaches thereby ensuring the most appropriate and robust evidence informs decision-making. Collaboration can lead to a more comprehensive view of a shock at hand and enable a faster assessment of potential impacts to infrastructure, human lives, economic, culture and the environment. Similarly, it facilitates exchange of information and better access to data, evidence or expertise providing a rich picture of what is happening on the ground, enabling a hands-on view of how a system works, and bringing together practical expertise and theoretical knowledge, enhancing overall quality of data, relationships and responses. Overall, when managed well, collaboration can enable a better pooling of resources to increase the effectiveness in managing and responding to flooding or heatwave events particularly where communities are more resilient and are therefore better able to support the vulnerable in their communities.

Whilst a wide range of evidence exists on the implementation of decision-making processes, in the context of nexus shocks, these are

Table 5.1   Key recommendations for improving resilience to climate shocks

Responses	• *Contextualization of warnings* in relation to recent events is useful approach to help people understand the scale of what they are experiencing and to trigger action to reduce impacts
	• *Move away from abstraction of warnings to situational warnings,* which provide a clear message on the situation and required response
	• *Clarity on what expected impact and action required are.* It is important to convey how events are expected to impact recipients, to make it relevant so the communication is understood
	• *Pro-active and pre-emptive approaches* for shock events with the bigger picture in mind
Governance	• *Importance of having structures in place to share and disseminate effectively* through a range of communication channels ensuring correct information is feeding through planning, preparation, responding
	• *Credibility and trust essential.* Lack of trust in organization can undermine messaging
Communication	• *Digestible and targeted information to audience is needed,* ensuring requirements are being met through effective communication with the clear purpose of the message, through clear, simple and effective messaging
	• *Need a common language and operating picture* to improve consistency of message
	• *Package information differently* and preference to avoid information in order to avoid uncertainty of need for action
	• Working with growth in technology for communication during emergency events
Co-production	• *Relationships need to be built on trust and understanding* of the needs of others'
	• *Networks should be established* to enable cross-stakeholder collaboration and knowledge exchange around key responses to weather and climate extremes whilst building capacity and knowledge of the public and members of those networks
	• *Have clarity of roles,* mechanisms of working and remits of stakeholders engaged in decision-making process
	• Better systems needed to *recognise the value of collaborative research* where scientists are recognized for their work and the broader research question and its practical value are acknowledged
	• *Flexibility and agile processes* needed where contributors are open to doing things differently and sharing, innovatively and in a way that aligns with the needs of communities affected by nexus shocks
	• *More active collaboration between policy/practitioners and academic* where they are viewed as partners rather than stakeholders and therefore would have more ability to influence

(continued)

Table 5.1   (continued)

Support	• Put in place system to *maintain historical and institutional memory* so that when shock events re-occur, that knowledge and expertise on how to respond is not lost • *Provide funding mechanisms* whereby policy and practitioners actors are eligible for funding and have more leverage over how Research Councils may spend their money so that it can be more applied • *Improved constructive conversations* where people talk to each other to stimulate better frameworks for collaboration and accounting for benefits of collaborative projects

less well understood, particularly when considering the impacts and implications on sectors and stakeholders working across the food, energy, water, environment sectors. Improving understanding of how decision-making processes are affected by climate and weather-related shocks can be improved in a number of ways:

- *Embedding reflection within decision-making processes* on the different roles and responsibilities held by those involved in such processes, those affected by nexus shocks and those managing and responding to shocks such as heatwaves and flooding. By doing this, a more iterative approach can be adopted whereby better consideration is given for the impacts of decisions as a result of nexus shocks, and a clearer understanding of how decisions are affected by availability of evidence, flows of information, cultures of responses, adaptive capacity and collaborative processes.
- Different responses may occur within a single shock event which may be underpinned by different sources of evidence (whether these be meteorological data, modelling forecasts, risk assessments, qualitative data and so forth). This therefore calls for a *thorough examination of evidence used by different stakeholders*, where this is sourced, how well this meets end-user needs and the extent to which it contributes to or conflicts with other existing resources (in the UK, for example, this could be the Climate Change Risk Assessment).
- In addition to evidence sources, scientists and academics are not always *packaging their findings* in ways to successfully reach practitioners; efforts are underway to improve this, but there is insufficient focus on the language. There are differing views by

stakeholders on what is considered good practice in effective dissemination of evidence, collaboration with other stakeholders, and the extent to which they are involved in co-production, whether this is perceived to be a 'good approach', and identifying lessons learnt in communicating uncertainty and risk. In identifying the extent to which stakeholders collaborate, with whom, the perceived benefits and challenges that emerge from this, and what works and doesn't work well in collaborative processes can shed light into tensions that may exist within decision-making processes, conflicting priorities, contextual influences and the adaptive capacity of an area subjected to nexus shocks.

- Adopting a *process of co-production involving end-users* enables access to valuable knowledge (such as community knowledge on the location of vulnerable people) that may otherwise be untapped. It therefore enables collaboration stakeholders together but also funders to help co-develop innovative, agile and responsive modes of funding that are more suitable to the current context of climate shocks and those working across the nexus who are impacted by these shocks and will need to develop resilient responses. A limitation of co-production processes is that they are considered to be time and resource intensive, particularly to engage with and maintain relationships with sustained trust and willingness on end-users part to engage. Whilst this tends to lead to very positive experiences, apathy to engage can exist, calling for more capacity, resources and realignment of timescales.

- *Capturing failures, successes and lessons learnt* provide useful ways of building a portfolio of evidence in addition to data that will underpin decisions on resilience to climate shocks. In particular these tend to show that efforts should be focused primarily to deliver good, needs-based evidence; to develop better, targeted communication; that collaboration plays an important role; that people should be an important focus of responses; how to better deal with uncertainty; and develop proactive, integrated responses with strong leadership and governance structures.

- Delivering resilience responses to climate shocks requires an *alignment in the priorities and deployment of resources of policymakers, practitioners, the science/academic* community as well as funders. This primarily enables the evolution of decision-making support by building collaboration and strengthening relationships and an evidence base; focusing on more applied practical application of solutions and facilitating better communication and building capacity to address the issue. Research funders have an important role to play here to support this, among others, by allocating more resources to applied, practical projects, with a greater focus on co-production and collaboration which involves end-users in the shaping of research funding programmes at the highest of levels early on.

CONCLUSION

Communication, collaboration and co-production are elementary to build resilience in response to climate shocks, particularly when these affect the food, energy, water, environment nexus. Mechanisms on how this can be effectively embedded within decision-making processes have been discussed in this book and reflect the views and concerns of those who participated in the Nexus Shocks project activities. A key challenge, however, is the availability of resource and the capacity available to support this in the longer term. Whilst there are widespread calls from policy, academia, practice and other sectors to facilitate better communication, collaboration and co-production across stakeholders and sectors and to the public, in order to enhance efforts to build resilience to shocks such as heatwaves as flooding, this is sparsely accompanied with financial resources to do so. The financial investment would be most effectively deployed in two ways, the first would be by *funding individual roles and dedicated teams* within organizations to champion this and implement processes that align with the guidance outlined in this book. This would be distinct from more traditional Communications team who often focus predominantly on Public Relations, engagement with the media and responses to press enquiries. These individuals would be in place specifically to implement and manage processes of communication, collaboration and co-production within and external to the organization in a nexus shock context. A second mechanism would be to *invest in capacity building* of these individuals and teams of individuals to ensure they are familiar with the most up to date methods, processes and research which underpins these to facilitate proper implementation. This would ultimately enable a closer alignment of the needs, interests and plans of all stakeholders involved in the event of a shock event and properly trained individuals to facilitate mechanisms of sound communication, collaboration and co-production.

There are multiple co-benefits to embedding communication, collaboration and co-production in resilience building to climate shocks. These will ultimately minimize impacts of these shocks to ever-growing and vulnerable populations whose vulnerability is likely to shift and increase under a changing climate. Communication, collaboration and co-production across disciplinary boundaries and sectors requires systemic and culture changes across academia, policy, practice and communities. Appropriate resources must be deployed in a flexible manner to ensure impacts and risks are addressed as swiftly as possible. Understanding of the ways in which other stakeholders operate under shock scenarios can enable a closer working relationship and alignment with cultures, values and beliefs of those involved in decision-making processes. Finally, mechanisms for communication are widespread, however with science underpinning effective decision-making, the scientific community, their institutions, funders and publishers must alter their own practices to ensure scientific advice truly delivers impact where it matters in the context of climate shocks to the food, energy, water, environment nexus.

REFERENCES

ASC. (2016). *UK climate change risk assessment evidence report 2016 method document, version 1.0.* London: Adaptation Sub-Committee, Committee on Climate Change.

Bammer, G. (2013). *Disciplining interdisciplinarity: Intergation and implementation sciences for researching complex real world problems* (472 pp.). Canberra: ANU Press.

Benzie, M. (2014). Social justice and adaptation in the UK. *Ecology and Society, 19*(1), 39.

Benzie, M., Harvey, A., Burningham, K., Hodgson, N., & Siddiqi, A. (2011). *Vulnerability to heatwaves and drought: Case studies of adaptation to climate change in South-West England.* York: Joseph Rowntree Foundation.

Bidwell, D., Dietz, T., & Scavia, D. (2013). Fostering knowledge networks for climate adaptation. *Nature Climate Change, 3*, 610–611.

Burchell, K., Fagan-Watson, B., King, M., Watson, T., Cooper, C., Holland, D., et al. (2017). *Developing the role of community groups in local climate resilience: Final report of the urban heat project.* London: Policy Studies Institute. psi.org.uk/urban_heat.

Cash, D., Clark, W., Alcock, F., Dickson, N., Eckley, N., & Jäger, J. (2002). *Salience, credibility, legitimacy and boundaries: Linking research, assessment and decision making* (Harvard University Faculty Research Working Papers Series). Cambridge, MA: Harvard University.

Christel, I., Hemment, D., Bojovica, D., Cucchiettia, F., Calvo, L., Stefaner, M., et al. (2017). Introducing design in the development of effective climate services. *Climate Services, 9,* 111–121. https://doi.org/10.1016/j.cliser.2017.06.002.

Clarke, B., Stocker, L., Coffey, B., Leith, P., Harvey, N., Baldwin, C., et al. (2013). Enhancing the knowledge governance interface: Coasts, climate and collaboration. *Ocean and Coastal Management, 86,* 88–99.

Collins, H. M., & Evans, R. J. (2002). The third wave of science studies: Studies of expertise and experience. *Social Studies of Science, 32*(2), 235–296.

Corburn, J. (2007). Community knowledge in environmental health science: Co-producing policy expertise. *Environmental Science & Policy, 10*(2), 150–161.

Cruickshank, J. (2017). Local hegemonies resisting a green shift and what to do about it: The introduction of a regional park in southern Norway. *Journal of Environmental Policy & Planning.* https://doi.org/10.1080/15239 08X.2017.1398640.

DFID. (2011). *Defining disaster resilience: A DFID approach paper.* London: Department for International Development.

Duit, A., & Galaz, V. (2008). Governance and complexity—Emerging issues for governance theory. *Governance, 21*(3), 311–335.

Durose, C., Beebeejaun, Y., Rees, J., Richardson, J., & Richardson, L. (2011). *Towards co-production in research with communities.* Swindon: AHRC. Available online at: http://www.ahrc.ac.uk/documents/project-reports-and-reviews/connected-communities/towards-co-production-in-research-with-communities/.

EU. (2016). *RESIN Project Fact Sheet.* European Union Horizon 2020. Accessed 27 February 2018. Available at: http://www.resin-cities.eu/fileadmin/user_upload/CORDIS_project_196890_en.pdf.

Fischer, F. (2000). *Citizens, experts, and the environment: The politics of local knowledge.* Durham: Duke University.

Howarth, C. (2016). *Responding to extreme weather events* (ESRC Evidence Briefing).

Howarth, C., & Monasterolo, I. (2016). Understanding barriers to decision-making in the UK energy-food-water nexus: The added value of interdisciplinary approaches. *Environmental Science & Policy, 61,* 53–60.

Howarth, C., & Monasterolo, I. (2017). Opportunities for knowledge co-production across the energy-food-water nexus: Making interdisciplinary approaches work for better climate decision making. *Environmental Science & Policy, 75,* 103–110.

Ingold, K., & Fischer, M. (2014). Drivers of collaboration to mitigate climate change: An illustration of Swiss climate policy over 15 years. *Global Environmental Change, 24,* 88–98.

Jasanoff, S. (2010). A new climate for society. *Theory, Culture and Society, 27*(2–3), 233–253.

Johnson, N., & McGuinness, M. (2016). Flood resilience in the context of shifting patterns of risk, complexity and governance: An exploratory case study. In *E3S FLOOD Risk 2016—3rd European Conference on Flood Risk Management*.

Lefevre, C., Bruine de Bruin, W., Taylor, A., Dessai, S., Kovats, S., & Fischhoff, B. (2015). Heat protection behaviours and positive affect about heat during the 2013 heat wave in the United Kingdom. *Social Science and Medicine, 128*, 282–289.

Löhr, K., Weinhardt, M., Graef, F., & Sieber, S. (2017). Enhancing communication and collaboration in collaborative projects through conflict prevention and management systems. *Organizational Dynamics.* https://doi.org/10.1016/j.orgdyn.2017.10.004.

Martin, S. (2010). Co-production of social research: Strategies for engaged scholarship. *Public Money & Management, 30*(4), 211–218.

Moezzi, M., & Janda, K. B. (2014, March). From "if only" to "social potential" in schemes to reduce building energy use. *Energy Research and Social Science, 1*, 30–40. http://dx.doi.org/10.1016/j.erss.2014.03.014.

Newton, J. (2017). Shared intentions: The evolution of collaboration. *Games and Economic Behavior, 104*, 517–534.

Ostrom, E. (1996). Crossing the great divide: Co-production, synergy and development. *World Development, 24*(6), 1073–1088.

Pelling, M., & Dill, K. (2010). Disaster politics: Tipping points for change in the adaption of socio-political regimes. *Progress in Human Geography, 34*(1), 21–37.

Polk, L. (2015). Transdisciplinary coproduction: Designing and testing a transdisciplinary research framework for societal problem solving. *Futures, 65*, 110–122.

Scandelius, C., & Cohen, G. (2016). Sustainability program brands: Platforms for collaboration and co-creation. *Industrial Marketing Management, 57*, 166–176.

Smith, S., Elliot, A., Hajat, S., Bone, A., Smith, G., & Kovats, S. (2016). Estimating the burden of heat illness in England during the 2013 summer heatwave using syndromic surveillance. *Journal of Epidemiology and Community Health, 1*, 1–7.

Stilgoe, J., & Wilsdon, J. (2014). Why should we promote public engagement with science? *Public Understanding of Science, 23*(1), 4–15.

Tabara, J. D., St. Clair, A. L., Hermansen, E. A. T. (2017). Transforming communication and knowledge production processes to address high-end climate change. *Environmental Science & Policy, 70*, 31–37.

Turney, J. (2014). To be inclusive, you need more voices. Q&A with Sheila Jasanoff. *Future Earth Blog.* Last accessed 13 October 2015. Available online at: http://www.futureearth.org/blog/2014-jul-23/be-inclusive-you-need-more-voices-qa-sheila-jasanoff.

Twigger-Ross, C., Brooks, K., Papadopoulou, L., & Orr, P. (2015). *Community resilience to climate change: An evidence review*. York: Joseph Rowntree Foundation.

Voorberg, W. H., Bekkers, V. J. J. M., & Tummers, L. G. (2014). A systematic review of co-creation and co-production: Embarking on the social innovation journey. *Public Management Review*. https://doi.org/10.1080/14719037.2014.930505.

Webber, S. (2017). Circulating climate services: Commercializing science for climate change adaptation in Pacific Islands. *Geoforum, 85,* 82–91.

Wentworth, J. (2016). *The water-energy-food nexus*, POSTNOTE, 543. London: Parliamentary Office of Science and Technology.

Wolf, T., Gosling, S., McGregor, G., & Pelling, M. (2007). *The social impacts of heat waves*. London: Environment Agency.

Index

© The Editor(s) (if applicable) and The Author(s) 2019
C. Howarth, *Resilience to Climate Change*,
https://doi.org/10.1007/978-3-319-94691-7